AN INTRODUCTION TO THE MATHEMATICS OF PLANNING AND SCHEDULING

Geza Paul Bottlik

NEW YORK AND LONDON

First published 2017
by Routledge
711 Third Avenue, New York, NY 10017

and by Routledge
2 Park Square, Milton Park, Abingdon, Oxon OX14 4RN

Routledge is an imprint of the Taylor & Francis Group, an informa business

Library of Congress Cataloging-in-Publication Data
Names: Bottlik, Geza Paul, author.
Title: An introduction to the mathematics of planning and scheduling / Geza Paul Bottlik.
Description: New York, NY : Routledge, [2016] | Includes bibliographical references and indices.
Identifiers: LCCN 2016043141 | ISBN 9781482259216 (hbk) | ISBN 9781138197299 (pbk) | ISBN 9781315381473 (ebk) | ISBN 9781315321363 (mobi/kindle) | ISBN 9781482259254 (web PDF) | ISBN 9781482259278 (ePub)
Subjects: LCSH: Production scheduling—Mathematics. | Production planning—Mathematics.
Classification: LCC TS157.5 .B675 2016 | DDC 658.5001/51—dc23
LC record available at https://lccn.loc.gov/2016043141

ISBN: 978-1-482-25921-6 (hbk)
ISBN: 978-1-138-19729-9 (pbk)
ISBN: 978-1-315-38147-3 (ebk)

Typeset in Bembo and Myriad
by Apex CoVantage, LLC

CONTENTS

FIGURES

PREFACE

This text is intended as a one-semester course for first-year graduate students in Industrial and Systems Engineering or Operations Management, some with bachelor's degrees from other areas. I have been teaching such a course at the University of Southern California for 26 years. I have also taught it at the University of Michigan for a couple of years. I have experimented with a number of books during that time and have found each that I have tried to be useful, but only in a limited area. None seemed to be suited as a single work to cover the whole semester and be. at a level for first-year graduate students. I have used substantial portions of Simon French's text that I found to be appropriate, but unfortunately it is out of print and also somewhat outdated, having been last revised in 1982. I am much indebted to Dr. French, both for the many years of teaching that I got out of his text and his generous permission to modify and use much of his text and for the use of many of his examples.

The intent of this course is principally to acquaint students with the necessity for scheduling activities under conditions of limited resources in industrial, service, and public and private environments and to introduce them to methods of solving these problems. In order to put this material in perspective, I have been introducing the class to production planning before tackling the details of scheduling. For this purpose, I have again used a number of different texts, most of the time the one by Vollman. This is an excellent text for a comprehensive class in operations management, but was too extensive and somewhat too detailed for my purposes. I feel that I need material that is more focused as an introduction to scheduling.

I have come to the conclusion that the many students taking this class would benefit by having a concise text that covered the material in the right amount of detail.

This is that text.

The usual length of this class is 15 weeks, but on occasion we do a summer session for only 12 weeks. In that case I skip the more specialized material contained in Chapters 14 and 15 on relaxations and on stochastic problems as well as references to project scheduling in the Appendix.

For more information on this text, visit www.gezabottlik.com/Intromath PlanSched.html.

PREFACE

1 INTRODUCTION

There is little doubt in my mind that human beings have been scheduling their activities for tens of thousands of years. After all, one had to decide whether to go looking for mammoth or to have breakfast first. It is highly unlikely, on the other hand, that any mathematical approaches were involved in that exercise. But some logic undoubtedly was. And not much has changed for many people since—most individual and many service decisions, and even some industrial ones, are made by people on the fly and purely by intuition, mostly based on prior experience. However, there has also been extensive progress in developing both a theory and practical approaches to solving the problem of providing goods and services in a planned and affordable manner that recognizes the many constraints that are imposed on these solutions.

When you were getting on an airplane, did you ever wonder why that particular airplane? Why that crew? Why was it on time or delayed? Along the same vein—how did that box of cereal that you are about to pull off the supermarket shelf get there? All around us, all day long, we encounter objects and services that we simply use, generally not thinking about how they became available. Planning and scheduling is the technique that ensures that the objects and services happen as they are supposed to (well, most of the time).

In this text we will explore how companies plan their work and services and how they schedule them in detail. While our main topic is the scheduling part, we also need to understand the context in which it occurs—the original plans and the associated costs and inventories. Planning and the underlying software methods are in Chapters 4 and 5, a brief introduction to inventory is in Chapter 3, and an overview of costs is covered in the Appendix. The remainder of the book is dedicated to scheduling, and we begin with an explanation of scheduling.

An Introductory Example

Four older gentlemen share an apartment in Los Angeles. Albert, Bertrand, Charles, and Daniel have not given up on newspapers despite the not so recent advent of the Internet. Their interests are quite varied and they are rather set in their preferences. Each Sunday they have four newspapers delivered: the *Financial Times*, the *Los Angeles Times*, the *Enquirer*, and the *New York Times*. Being small-minded creatures of habit, each member of the apartment insists on reading all the papers in his own particular order. Albert likes to begin with the *Financial Times* for 1 hour. Then he turns to the *Los Angeles Times* taking 30 minutes, glances at the *Enquirer* for 2 minutes, and finishes with 5 minutes spent on the *New York Times*. Bertrand prefers to begin with the *Los Angeles Times* taking 1 hour 15 minutes; then he reads the *Enquirer* for 3 minutes, the *Financial Times* for 25 minutes, and the *New York Times* for 10 minutes. Charles begins by reading the *Enquirer* for 5 minutes and follows this with the *Los Angeles Times* for 15 minutes, the *Financial Times* for 10 minutes, and the *New York Times* for 30 minutes. Finally, Daniel starts with the *New York Times* taking 1 hour 30 minutes, before spending 1 minute each on the *Financial Times*, the *Los Angeles Times*, and the *Enquirer* in that order. Each is so insistent upon his particular reading order that he will wait for his next paper to be free rather than select another. Moreover, no one will release a paper until he has finished it. Given that Albert gets up at 8:30 A.M., Bertrand and Charles at 8:45 A.M., and Daniel at 9:30 A.M., and that they can manage to shower, shave, dress, and eat breakfast while reading a newspaper, and given that each insists on reading all the newspapers before going out, what is the earliest time that the four of them can leave together for a walk in the park?

The problem faced by Albert and his friends, namely in what order they should rotate the papers among themselves so that all the reading is finished as soon as possible, is typical of the scheduling problems that we will be considering. Before describing the general structure of these problems and giving some examples that are, perhaps, more relevant to our modern society, it is worth examining this example further.

The data are rewritten in more compact form in Figure 1.1. How might you tackle this problem? Perhaps it will be easiest to begin by explaining what is meant by a reading schedule. It is a prescription of the order in which the papers rotate between readers. For instance, one possible schedule is shown in Figure 1.2, where

Reader	Gets up at	Reading Order and Times in Minutes			
Albert	8:30	F	L	E	N
		60	30	2	5
Bertrand	8:45	L	E	F	N
		75	3	25	10
Charles	8:45	E	L	F	N
		5	15	10	30
Daniel	9:30	N	F	L	E
		90	1	1	1

Figure 1.1 The Data for the Newspaper Reading Problem

	Read by			
Paper	1st	2nd	3rd	4th
F	A	D	C	B
L	B	C	A	D
E	C	B	A	D
N	D	A	C	B

Figure 1.2 A Possible Reading Schedule

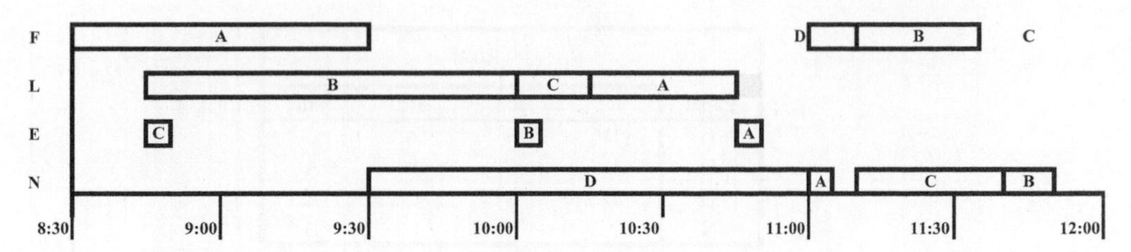

Figure 1.3 Gantt Diagram for the Schedule in Figure 1.2

A, B, C, and D denote Albert, Bertrand, Charles, and Daniel respectively. Thus Albert has the *Financial Times* (F) first, before it passes to Daniel, then to Charles, and finally to Bertrand. Similarly the *Los Angeles Times* (L) passes between Bertrand, Charles, Albert, and Daniel in that order. And so on.

We can work out how long this reading schedule will take by plotting a simple diagram called a Gantt chart (see Figure 1.3). In this we plot four time axes, one for each newspaper. Blocks are placed above the axes to indicate when and by whom particular papers are read. For instance, the block in the top left hand corner indicates that Albert reads the *Financial Times* from 8:30 to 9:30. To draw this diagram we have rotated the papers in the order given by Figure 1.2 with the restriction that each reader follows his desired reading order. This restriction means that for some of the time papers are left unread, even when there are people who are free and have not read them yet; they must remain unread until someone is ready to read them next. For instance, Bertrand could have the *Financial Times* at 10:00 A.M., but he wants the *Enquirer* first and so leaves the *Financial Times.* Similarly the schedule is also responsible for idle time of the readers. Between 10:15 and 11:01 Charles waits for the *Financial Times*, which for all but the last minute is not being read, but Charles cannot have the paper until after Daniel because of the schedule.

From the Gantt diagram, you can see that the earliest that all four can go out together is 11:51 A.M. if they use this schedule. So the next question facing us is: can we find them a better schedule, i.e., one that allows them to go out earlier? This question is left for you to consider in the first set of problems. However, before attempting those, we should consider what we mean by feasible and infeasible schedules.

In Figure 1.2 you were simply given a schedule without any explanation where it came from and we saw in the Gantt diagram that this schedule would work; it is possible for Albert and his roommates to pass the papers among themselves in this order. But suppose you were given the schedule that is shown in Figure 1.4. This schedule will not work. Albert is offered the *New York Times* first, but he does not want it until he has read the other three papers. He cannot have any of these until Daniel has finished with them and Daniel will not start them until he has read the *New York Times*, which he cannot have until Albert has read it.

In scheduling theory the reading orders given in Figure 1.1 are called the technological constraints. The common industrial terms for these are routing or processing order or process plan and are dictated by the characteristics of the processes

Paper	Read by			
	1st	2nd	3rd	4th
F	D	B	A	C
L	D	C	B	A
E	D	B	C	A
N	A	D	C	B

Figure 1.4 An Infeasible Schedule

involved. Any schedule that is compatible with these is called feasible. Thus Figure 1.2 gives a feasible schedule. Infeasible schedules, such as that in Figure 1.4, are incompatible with the technological constraints. Obviously, to be acceptable a solution to a scheduling problem must be feasible.

Problems

Please try these problems now before reading any further. It is true that you have been given little or no guidance on how they are to be solved. This was done for a good reason. Scheduling is a subject in which the problems tend to look easy, if not trivial. They are, on the contrary, among the hardest in mathematics. You will not appreciate this without trying some for yourself. Solving them is relatively unimportant; we'll solve them for you shortly anyway. What is important is that you should discover their difficulty.

1. Is the schedule in Figure 1.5 feasible for Albert and his friends?
2. How many different schedules, feasible or infeasible, are there?
3. What is the earliest time that Albert and his friends can leave for the park?
4. Daniel decides that the pleasure of a walk in the park is not for him today. Instead he will spend the morning in bed. Only when the others have left will he get up and read the papers. What is the earliest time that Albert, Bertrand, and Charles could leave?
5. Whether or not you have solved Problems 3 and 4, consider how you would recognize the earliest possible departure time. Do you need to compare it explicitly with those of all the other feasible schedules, or can you tell without this process of complete enumeration of all the possibilities?

Paper	Read by			
	1st	2nd	3rd	4th
F	C	D	B	A
L	B	C	A	D
E	B	C	A	D
N	B	A	D	C

Figure 1.5 Is This Schedule Feasible?

Albert, Bertrand, Charles, and Daniel's Apartment Revisited

Please do not read this section until you have tried the above problems.

Problems like that concerning Albert and his friends are so important to the development of the theory that we should pause and examine their solution in some detail.

1. *Is the given schedule feasible?* The short answer is no, and we may discover this in a number of ways. We might try to draw a Gantt diagram of the schedule and find that it is impossible to place some of the blocks without conflicting with either the schedule or the technological constraints. Alternatively we might produce an argument similar to, but more involved, than that by which we showed the schedule in Figure 1.4 to be infeasible. You will probably agree, though, that neither of these methods is particularly straightforward and, moreover, that the thought of extending either to larger problems is awesome. What we need is a simple scheme for checking the schedule operation by operation until either a conflict with the technological constraints is found or the whole schedule is shown to be feasible. The following illustrates such a scheme.

First we write the schedule and the technological constraints side by side as in Figure 1.6. We imagine that we are operating the schedule. We shall assign papers to the readers as instructed by the schedule. As the papers are read we shall mark the operations to indicate that they are completed and pass the papers to their next readers. Either we shall meet an impasse or we shall show that the schedule is feasible. We label the marks in the order in which they are entered.

We begin in the top left hand corner of the schedule. Charles is given the *Financial Times,* but will not read it until he has read the *Enquirer* and the *Los Angeles Times.* So we must leave this operation uncompleted and unmarked. Proceeding down the schedule, Bertrand is given the *Los Angeles Times* and we see from the technological constraints that he is immediately ready to read it. So we mark this operation both in the schedule and in the technological constraints. Next Bertrand is given the *Enquirer* and we see that, now he has read the *Los Angeles Times,* he is immediately ready to read it. So this is the second operation to be marked. We see that The *New York Times* is also assigned to Bertrand, but he is not ready to read it, so we leave this operation unmarked. We continue by returning to the top line of the schedule and by repeatedly working down the schedule, checking the leftmost unmarked operation in each line to see if it may be performed. Thus we show

Schedule from Figure 1- 5				
	Read by			
Paper	1st	2nd	3rd	4th
F	C	D	B	A
L	B	C	A	D
E	B	C	A	D
N	B	A	D	C

TechnologicalConstraint				
Reader	Reading Order			
A	F	L	E	N
B	L	E	F	N
C	E	L	F	N
D	N	F	L	E

Figure 1.6 Schedule and Technological Constraint

	Schedule from Figure 1-5			
	Read by			
Paper	1st	2nd	3rd	4th
F	**C 5**	D	B	A
L	**B 1**	**C 4**	A	D
E	**B 2**	**C 3**	A	D
N	B	A	D	C

	Technological Constraint			
Reader	Reading Order			
A	F	L	E	N
B	**L1**	**E2**	F	N
C	**E3**	**L4**	**F5**	N
D	N	F	L	E

Figure 1.7 Impasse in Completing the Schedule

that Charles may read the *Enquirer*, the *Los Angeles Times*, and the *Financial Times* without any conflict with the technological constraints. The position is now shown in Figure 1.7 with five operations outlined and we can see that an impasse has been reached. Each paper is to be assigned to a reader who does not wish to read it yet. Hence the schedule is infeasible.

We won't have to check for feasibility very often because the algorithms and methods that we will study are designed so that they cannot produce infeasible schedules. However, when creating schedules for large problems, as we will do in Chapter 7, we will need an explicit method. If you need to check a schedule for a more conventional problem based upon jobs and machines, you should have no difficulty in translating the method from the present context; just remember that here Albert and friends are the jobs, while the papers are the machines.

2. *How many different schedules, feasible or infeasible, are there?* If we let m = the number of machines and n = the number of jobs then a schedule for this problem consists of m permutations of the n jobs. Each permutation gives the processing sequence of jobs on a particular machine. Now there are $n!$ different permutations of n objects and, because each of the m permutations may be as different as we please from the rest, it follows that the total number of schedules is $(n!)^m$. In the problem facing Albert and his friends $n = 4$ and $m = 4$. So the total number of schedules is $(4!)^4 = 331{,}776$. Of these there are only 14,268 feasible ones and actually two optimum ones (found by complete enumeration). It is also noteworthy to note that the worst schedule takes more than twice as long as the optimum one—so it does pay to find the good schedules!

Here we should pause to consider the implications of these rather startling numbers. Here we have a very small problem: only four 'machines' and four 'jobs.' Yet the number of possible contenders for the solution is quite large. We can of course solve the problem by the simple expedient of listing all the possible schedules, eliminating the infeasible, and selecting the best of those remaining. But for most industrial problems this is not very practical, even considering today's very fast computers. Or tomorrow's for that matter. The size of these problems grows very rapidly. Consider that a guest was staying in the apartment so that there were five readers. The number of schedules would now be $(5!)^4 = 2.1 \times 10^8$! And a computer would take 625 times longer for this new problem! The very size of these numbers indicates the very great difficulty of scheduling problems. As you can surmise, real problems involve thousands of jobs on dozens of machines. To have any chance at all of solving them we must use subtlety. But even with the most subtle methods available we shall discover that some problems defy practical solution; to solve them would literally take centuries.

3. *What is the earliest time at which Albert and his friends may leave for the park?* Perhaps the easiest way for us to approach this problem is to look back at the schedule given in Figure 1.2 and see if we can improve upon it in any obvious way. Looking at the Gantt diagram (Figure 1.3) and, in particular the row for the *New York Times*, we see that it is the *New York Times* that is finished last. Moreover, it is left unread between 11:05 when Albert finishes it and 11:11 when Charles is ready for it. Thus 6 minutes are apparently wasted. Is there another schedule that does not waste this time, one that ensures that the *New York Times* is read continuously? Well yes, there is. Consider the schedule in Figure 1.8. This has the Gantt diagram shown in Figure 1.9.

Note that now the *New York Times* is read continuously and that it is the last paper to be finished. Thus under this schedule the earliest time at which they can leave for the park is 11:45. Moreover, some thought will convince most people that this is an optimal schedule. Everybody starts reading as soon as they can and once started the *New York Times* is read without interruption. There seems to be no slack left in the system. But there is. Consider the schedule in Figure 1.10.

	Read by			
Paper	1st	2nd	3rd	4th
F	A	B	C	D
L	B	C	A	D
E	C	B	A	D
N	D	A	C	B

Figure 1.8 An Improved Schedule

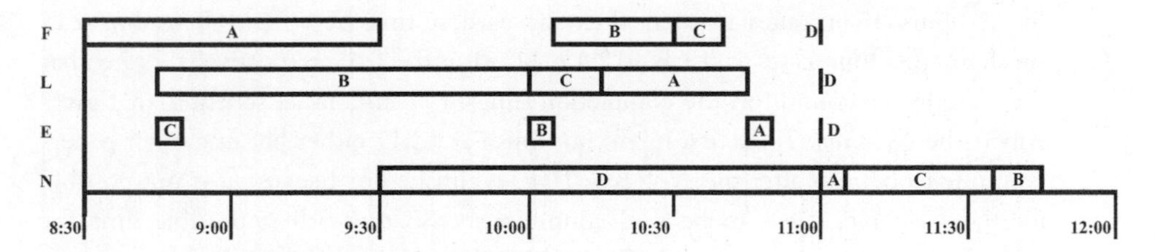

Figure 1.9 The Gantt Chart for the Schedule in Figure 1.8

	Read by			
Paper	1st	2nd	3rd	4th
F	C	A	B	D
L	C	B	A	D
E	C	B	A	D
N	C	D	B	A

Figure 1.10 Optimal Schedule

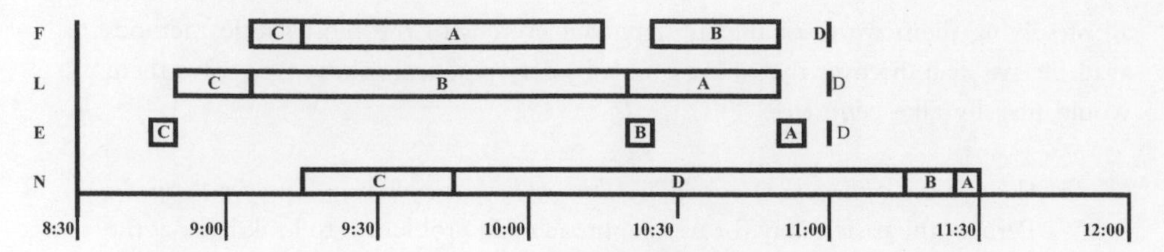

Figure 1.11 Gantt Diagram for the Optimal Schedule

This schedule leads to the Gantt diagram shown in Figure 1.11 and we see that all reading is now completed by 11:30, 15 minutes earlier than allowed by the schedule in Figure 1.8. So that schedule was clearly not optimal. How has this improvement been achieved?

Compare the rows for the *New York Times* in the two Gantt diagrams (Figures 1.9 and Figure 1.11). What we have done is 'leap-frogged' the block for Charles over those for Daniel and Albert. Because Charles can be ready for the *New York Times* at 9:15, if he is allowed the other papers as he wants them, we gain 15 minutes. Moreover, it is possible to schedule the other readers, Albert, Bertrand, and Daniel, so that this gain is not lost. The moral of all this is that in scheduling you often gain overall by not starting a job on a machine as soon as you might. Here Albert and Bertrand wait for the *Financial Times* and *Los Angeles Times* respectively. They could snatch up these papers as soon as they get up, but their patience is rewarded.

It turns out that the schedule in Figure 1.10 is optimal; no other schedule allows them to leave the apartment earlier. To see this we consider four mutually exclusive possibilities: Albert reads the *New York Times* before anyone else; Bertrand does; Charles does; or Daniel does. At the earliest Albert can be ready to read the *New York Times* at 10:02. (Check this from Figure 1.1.) The earliest times at which Bertrand, Charles, and Daniel can be ready are 10:28, 9:15, and 9:30 respectively. Thus if we assume that, once started, the *New York Times* is read continuously taking 2 hours 15 minutes in total, then the earliest time at which all reading can finish in the four cases is 12:17, 12:43, 11:30 and 11:45 respectively. Note that these are lower bounds on the completion times. For instance, a schedule that gives Albert the *New York Times* first might not finish at 12:17 either because other papers continue to be read after the *New York Times* is finished or because it is not possible for the *New York Times* to be read continuously. So the earliest possible time for any schedule to finish is $\min\{12{:}17, 12{:}43, 11{:}30, 11{:}45\} = 11{:}30$. Figure 1.6 gives a schedule completing at 11:30; it must, therefore, be optimal.

The structure of the preceding argument deserves special emphasis, for it will be developed into a powerful solution technique known as branch and bound (see Chapter 10). We had a particular schedule that completed finally at a known time. To show that this schedule was optimal, we considered all possible schedules and divided them into four disjoint classes. We worked out for each class the earliest that any schedule within that class could complete, i.e., we found a lower bound appropriate to each of these classes. We then noted that our given schedule completed at the lowest of the lower bounds. Thus no other schedule could complete before it and so it had to be an optimal schedule.

4. *What is the earliest time at which Albert, Bertrand, and Charles may leave without Daniel?* At 11:03. I leave with you both the problem of finding a schedule to achieve this and the problem of showing such a schedule to be optimal. However, I will give you one hint. Use a bounding argument like that above except that you should consider who is first to read the *Los Angeles Times*, not the *New York Times*.
5. *How does one prove a schedule to be optimal? Need one resort to complete enumeration?* For the particular scheduling problem facing Albert and friends we now know that complete enumeration is unnecessary. However, the solution of Problems 1, 2, and 3 involved a certain amount of luck, or rather relied on knowing the answer before we started. No straightforward logical argument led to the schedule in Figure 1.10. I just produced it rather like a magician pulling a rabbit from a hat. Moreover, the bounding argument that I used to show optimality relied heavily on the structure of this particular problem, namely that the optimal schedule allowed the *New York Times* to be read continuously. (Why is this particular feature important to the argument?) In short, I have been able to solve this problem for you simply because I set it. So the question remains: in general, is it necessary to use complete enumeration to solve scheduling problems? The answer is the rest of this book.

2 A BRIEF HISTORY

This short chapter is intended to give you a brief glimpse how the knowledge and methods that are the subjects of this book are developed. A number of inventions were very influential in this development. Most of these were not intended to directly influence planning and scheduling. Many people, both well known and unknown contributed both small and large steps in the development. And some changes have become obsolete since they were first introduced. Because this book is about the mathematical and software side of producing products and services, we do not concern ourselves with many important topics such automation, robotics, the development of sequential production lines or manufacturing cells.

We will not go back beyond the 20th century, as the things that interest us the most all date from the beginning of that century. The first of the items is the Gantt chart. It is named after Henry Gantt (1861–1919). He was trained as an engineer and developed an interest in efficiency and scheduling. He developed the idea of representing tasks over time in a chart sometime after 1910. We still use it today, as you will find many instances in the book. The other development of that era was a scientific approach to improve efficiency advocated by Frederick W. Taylor (1856–1915) in 1911.

In 1947, George Dantzig (1914–2005), a mathematics professor, developed what is known as the simplex method, a structured approach to find the optima of linear problems. His approach still forms the basis of almost all optimization methods.

The next invention that impacted our area is of course the computer. It's hardly necessary to say that computers have drastically changed our society, but here we are interested in their impact on planning. The first digital computers (Colossus and ENIAC) appeared during World War II. After the war their use expanded rapidly starting in the mid-1950s, especially once transistors replaced vacuum tubes. Computers that could be used routinely for business, especially banking, did not appear until the 1950s and were pretty much restricted to scientific experiments and banking.

Before computers were used in manufacturing, all the data were contained on index cards, generally one for each item on which were recorded all the relevant data for that part. Here it is necessary to jump ahead and give a definition of dependent and independent parts. Independent parts are those that are final products delivered to customers, such as cars or bicycles. Dependent items are those that are required to produce the independent items. Today we deal completely differently with each of these, but prior to about 1964 they were treated the same—each item, independent or dependent, was treated as if it was independent and an appropriate economic order quantity (EOQ) calculated for it (see Chapter 3). The timing for ordering and receiving each was also dealt with independently.

The great breakthrough for manufacturing planning came in 1964 with the introduction of Material Requirements Planning by John Orlicky (1922–1986). It was not until 1975 that Orlicky published a book describing the system. (The book has been republished several times much later, Ptak and Smith (2011).) The underlying concept was 'demand driven,' that is all dependent items were tied to the demand for the independent item. It made excellent use of the computer's capability of processing mountains of information in a relatively short time. Today the concepts sound so obvious, but they revolutionized the way in which manufacturing was conducted. Not only was it a game changer, but Orlicky's zeal in promoting the idea spread its use very rapidly.

The next big step came in 1983, with the expansion of Material Requirements Planning (also known thereafter as MRP I) into manufacturing requirements planning (MRP II) by Oliver W. Wight (1981). It expanded the use of computers into the areas of capacity planning, shop floor control, and other tasks having to do with operating a manufacturing operation. Since then both MRPs have been integrated into what is called ERP—enterprise requirements planning—which now encompasses all activities of an enterprise.

The ability to track both information and products was expensive and difficult to implement prior to the arrival of bar coding in 1974 in grocery stores. It had a long journey to commercialization from the first ideas by Bernard Silver and Norman Woodland in 1949. Its wide use in manufacturing took several more years, but today it is ubiquitous, spurred by the 1981 adoption of Code 39 by the US Department of Defense for all products.

A similar method of identification, radio frequency identification (RFID), is still expanding its uses after initial introduction in the 1970s.

The last major change started in the 1980s when the first scheduling programs appeared using finite scheduling as opposed to infinite capacity, fixed lead time scheduling. Most of these, as they are still today, are only capable of scheduling much shorter time periods than are required for planning. They have come to be known as APS (advanced production systems, advanced planning systems, or advanced planning and scheduling) and are expected to expand into having a larger role in planning as their and the computers' capabilities improve.

3 INVENTORY

Introduction

While few of us think of ourselves as managing inventory, in fact, we all do. Think about how you make sure that there are adequate groceries at home—you assess what and how much you need, how long it will take you to get it, and then proceed to purchase it. Or how about gas in your car—there is some sort of signal to indicate that you need it, hopefully not a sputtering engine, and usually there is adequate time to actually obtain it. This leaves you with the decision on how much to buy—because most of us have credit cards, we opt on filling up the tank. As a final example, take the issue of cash. How often and how much do we withdraw from an ATM? There are limits—how much money you have, how much the bank is willing to dispense each time, and when we will have an opportunity to access another ATM.

Now that you know that you are actually a practicing inventory manager, we can proceed to take a brief look at the basics of the mathematics behind formal inventory management.

The basic questions of inventory management are:

1. How should the status of the inventory be determined?
2. When should an order be placed?
3. How large should the order be?

These questions are answered by considering four factors:

1. The importance of the item;
2. How frequently the inventory is reviewed;
3. The selection of an inventory policy;
4. What our cost objective is.

Before we launch into definitions, it is important to realize that some things are bought to be used up, or resold, without any further processing. Most consumer goods are in this category at the retail and consumer level. Other items are bought to be processed, such as plastic pellets for molding parts, or onions to be used in a soup.

We will now start with a few definitions that will be helpful in understanding the ideas behind the formulas:

Replenishment quantity, usually designated as Q, is the amount you order at a given time. An example would be the two quarts of milk you buy each Saturday. The optimum economic replenishment quantity, Q* or EOQ, can be determined for most items, as we shall see later.

Lead time, usually designated as L, is the amount of time that has to elapse between the time that you recognize that something is needed and when you actually can get it.

Order point, OP or s, is the quantity remaining when you decide that it is time to order more. For example, when the little light on your car's dashboard comes on, there are about two gallons of gas left and it is time to look for a gas station.

Order up to level, designated as S, is the maximum quantity that you would want or can have at any given time. Good examples are the size of your gas tank or the space on the designated shelf for an item.

Raw material is material that has been purchased but has yet to have work done on it.

Work in process, WIP, is the material on which we have started working, i.e., what we have added value to, but have not yet turned into a finished product.

Finished goods are product that has been completed but has not yet been shipped to a customer.

Demand, D, is the total quantity required per year. For example, a hospital operating room performs 2,000 operations per year, so would require 2,000 anesthesiologists' appearances, or a car factory manufactures 20,000 of a particular model per year. Demand is also separated into *independent demand*, such as the cars in the previous definition, and *dependent demand*, such as the 80,000 wheels required for those 20,000 cars.

We further distinguish inventory as *cycle inventory*. It exists because time elapses between successive orders as items are used up. Think of the gas in your tank between full and the level at which the light comes on and you replenish it.

Another category is *safety stock* (buffer inventory), designated by SS—the amount we keep on hand for unpredictable circumstances. We derive this amount from the variations we expect from the usual usage of an item. If you ordinarily consume one can of tuna a week, you would buy one can per week. However, just in case the store is out of tuna in any given week, you keep an extra can at home at all times.

Finally there is *anticipation inventory*—the stock that we build up for a time when the demand rate will exceed our capacity to produce at the same rate.

Inventory is both beneficial and a substantial expense. You cannot produce or sell anything without it—that is the good part. On the other hand when it sits around not doing anything, it can be quite costly. Many factors go into the cost of carrying inventory—space, insurance, pilferage, obsolescence, the cost of borrowing money to pay for it, and so forth. This can be up to about 25% of the value of the item over a year. This is referred to as the inventory carrying cost, k.

The Economic Order Model

A very simple and nevertheless powerful and idealized model of inventory is shown in Figure 3.1a. The angled lines represent a constant rate of usage, while the vertical lines represent the instant replacement of the quantity Q when you run out. From the geometry of triangles we can tell that the average inventory is Q/2. In Figure 3.1b you can see that if we halve the time between replenishments, the average inventory and the resultant carrying costs are also halved. So why not replenish even more frequently? The answer is that there is also a cost associated with each replenishment—placing the order, receiving it, storing it, and paying for it are some of the contributors to this cost and the total cost increases linearly with the number of replenishments.

We are now in a position to derive what is the simplest, most used, and oldest concept in inventory management—the economic order quantity (EOQ)—by minimizing the total cost over time and using our definitions from earlier in the chapter. We just need to define two more variables, the cost of the item, C, and A, the cost of ordering a replenishment.

$$\text{Total Cost} = DC + \frac{QkC}{2} + DA/Q \qquad \text{Equ. 3.1}$$

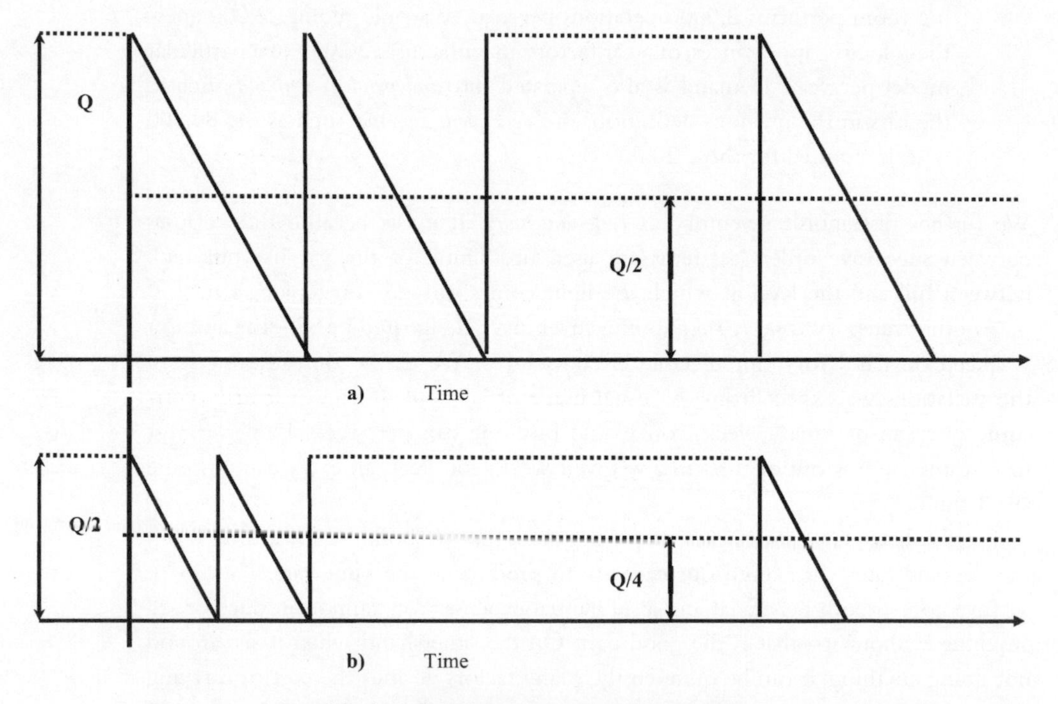

Figure 3.1 A Simplified Model of Inventory Quantity Over Time

The total cost consists of the direct cost of the item times the demand, plus the holding cost of the average inventory plus the cost of ordering. Note that we order D/Q times per year. When we differentiate this equation, set it equal to zero to obtain the minimum and solve for Q, we obtain:

$$EOQ = \sqrt{\frac{2AD}{kC}} \qquad \text{Equ. 3.2}$$

An example will demonstrate the reasons that this equation has and continues to dominate all discussions of inventory management. Suppose a store such as Costco sells 800 55-inch big-screen TVs per year. Each of these costs the store \$300. Each time they order a group of TVs, a cost of ordering of \$80 is incurred. Also, the inventory holding cost is 20% per year. The resulting EOQ is:

$$EOQ = \sqrt{\frac{2(80)(800)}{(20\%)(300)}} = 48.2 \qquad \text{Equ. 3.3}$$

Naturally, we cannot order 48.2 TVs. We can round to 50, which would mean ordering 800/50 = 16 times per year. It also might be more convenient to order every 4 weeks, or 800/(52/4), i.e., approximately 60 TVs at a time. The question is, how will that affect the total cost? To answer this, it is convenient to plot the two costs (note that the DC term is a constant and does not figure in our EOQ equation) as a function of quantity (Figure 3.2). You can see that the total cost curve is very shallow and as long as the selected quantity is reasonably close to the optimal quantity, the cost does not vary very much. This flexibility in choosing the quantity is very important—it allows us to accommodate requirements such as package sizes, minimum orders, truckloads, and similar limitations.

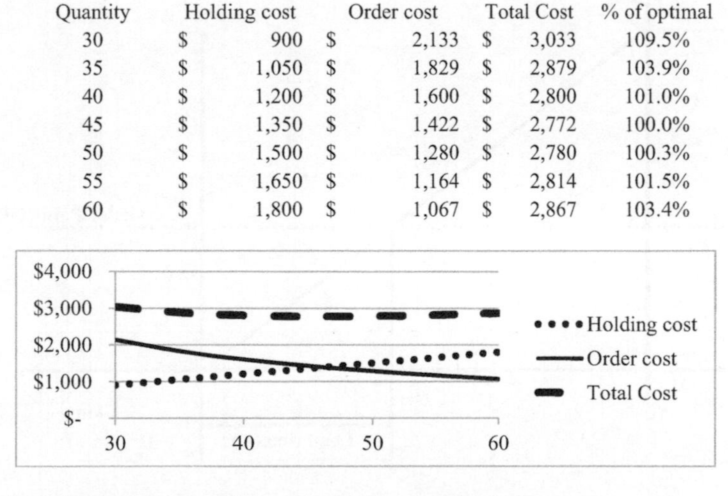

Quantity	Holding cost	Order cost	Total Cost	% of optimal
30	\$ 900	\$ 2,133	\$ 3,033	109.5%
35	\$ 1,050	\$ 1,829	\$ 2,879	103.9%
40	\$ 1,200	\$ 1,600	\$ 2,800	101.0%
45	\$ 1,350	\$ 1,422	\$ 2,772	100.0%
50	\$ 1,500	\$ 1,280	\$ 2,780	100.3%
55	\$ 1,650	\$ 1,164	\$ 2,814	101.5%
60	\$ 1,800	\$ 1,067	\$ 2,867	103.4%

Figure 3.2 Total Cost as a Function of Replenishment Quantity

Safety Stock

Our model thus far has been greatly simplified. It is time to add some refinements. The first of these is the realization that it is seldom possible to have instant replenishment when the item runs out. So we have to order before we run out. Or consumption continues while we wait for the replenishment to arrive. This is the lead time. We place our order when our supply reaches a certain point—this is the order point (OP) as shown in Figure 3.3.

The order point is determined by the lead time and the rate of consumption during that time:

$$OP = DL \quad \text{Equ. 3.4}$$

In our example we will assume that the supplier of TVs to Costco, perhaps Visio, quotes a lead time of 2 weeks. Our order point becomes 800(2/52) = 30.8. We usually would round this up to 31. How do we know that we have reached the order point? Do you constantly look at the dash board of your car to see if it is time to buy gas? Well, sort of—you glance there periodically. Fortunately, in most business we have computers, point of sale devices, and so forth to monitor our stock level and react when the order point is reached. We will examine the situations when this is not the case a little later, but first we have to also deal with uncertainties.

We have assumed a constant rate of usage D. Unfortunately, in reality this varies. If we assume a normal distribution and estimate the standard deviation, we can use this, coupled with a desired stock out rate, to determine the level of safety stock required. A stock out occurs when a customer asks for a product and we do not have any. The stock out rate is the percentage of ordering cycles that will experience a stock out. Let's use 5% so that the associated z level for 100% – 5% is 1.65.

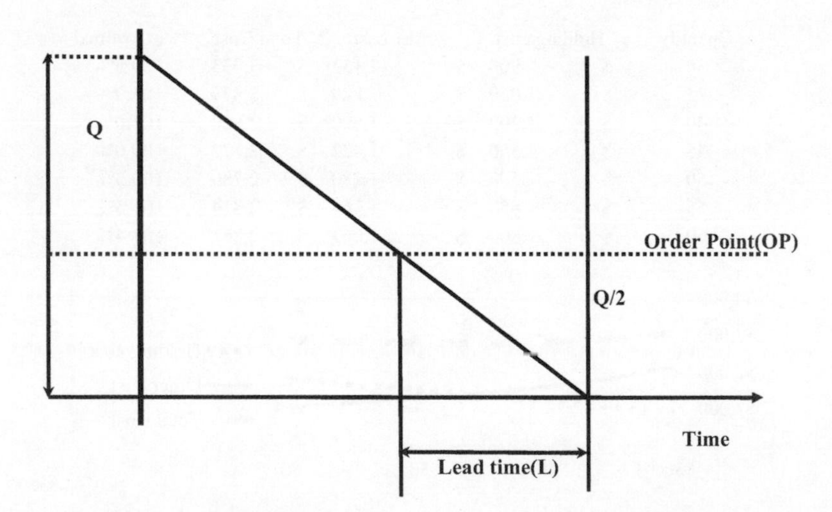

Figure 3.3 Order Point and Lead Time

Past experience has shown the retailer that the annual demand of 800 has a standard deviation $\sigma_D = 20$. But we are interested in the standard deviation only during lead time. The variation prior to reaching the order points is irrelevant to determining the safety stock.

$$\sigma_{DL} = \sigma_D\sqrt{L} \qquad \text{Equ. 3.5}$$

$$SS = z\,\sigma_{DL} \qquad \text{Equ. 3.6}$$

In our example $SS = 1.65(20)\sqrt{2/52} = 7$. This is the number of TVs we keep as insurance against running out. However, we can expect to have 5% of the cycles to experience a stock out. Because we have chosen to order every 4 weeks, or 13 times a year, our expected stockout will occur every 20/13 years, about every year and a half due to variations in demand, even with the protection of safety stock as shown in Figure 3.4. The increase in annual cost is 7(300)(20%) = \$420.

While suppliers promise a specific lead time, and frequently deliver within it, sometimes they do not. So it is useful to assume that the lead time is distributed normally and to know the standard deviation of the lead time, σ_L. We can use this information to increase our safety stock to also allow for this variation. Suppose that the supplier of the TVs indicates (or we glean from experience) that the standard deviation of the lead time is one day. Combining the variations (the bars indicate the average of lead time and demand during lead time):

$$\text{adjusted } \sigma = \sqrt{\sigma_{dl}^2\bar{L} + \sigma_L^2\overline{d^2}} = \sqrt{(16)2 + \frac{1}{7}\star(31^2)} = 8 \qquad \text{Equ. 3.7}$$

and our new safety stock is 1.65(8) = 14.

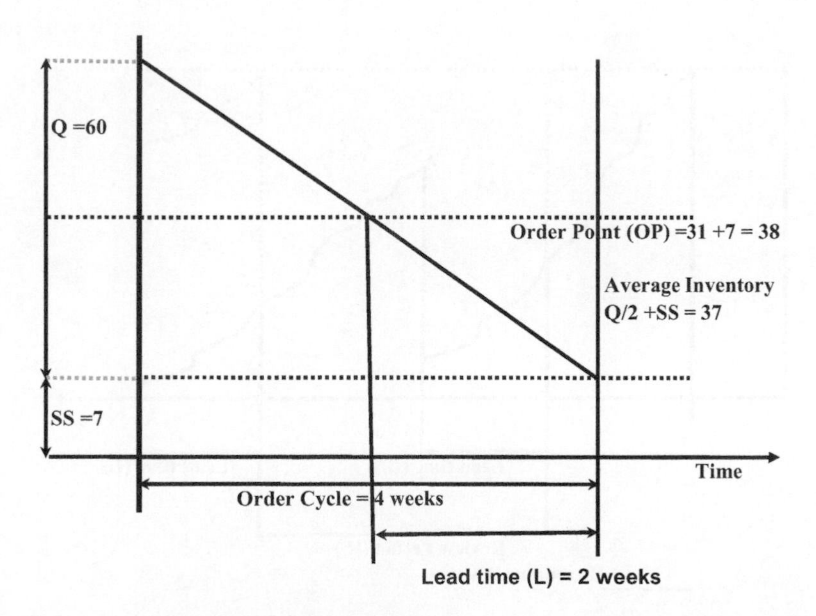

Figure 3.4 Safety Stock Added to Our Model

Period Review

So much for the case when the inventory level is being tracked for us. But suppose we can only check periodically, such as a weekly inventory assessment. We call this *period review* and refer to the elapsed time between reviews as R. This increases the period of uncertainty in the demand to the review period plus the lead time. This method also usually makes use of the idea of *order up to level*, S. We introduce one more concept, that of *inventory position*. Instead of only considering inventory already physically on hand, it adds any amount ordered but not yet received. See Figure 3.5. The wavy lines here indicate that the constant rate of usage is only an approximation.

$$\sigma_{DL} = \sigma_D \sqrt{L + R} \quad \text{Equ. 3.8}$$
$$R = Q/D \quad \text{Equ. 3.9}$$
$$\text{Demand during total lead time} = (R + L)\star D \quad \text{Equ. 3.10}$$
$$SS = z\sigma_{dL} \quad \text{Equ. 3.11}$$
$$S = SS + (R + L)\star D \quad \text{Equ. 3.12}$$

Continuing with our example, R = 60/800 = 0.075 or 4 weeks. σ_{DL} = 6.8 and SS = 12, and finally, S = 12 + (2 + 4)*800/52 = 105. Because the order quantity is usually different every time, it is difficult to estimate the inventory carrying cost without resorting to simulation.

We will look briefly at one more model—one that combines the review period with an order point. The effect of this is that if at the end of the review period, we only order if we are below the order point. Calculations for this model are only possible with trial and error. (See Figure 3.6.)

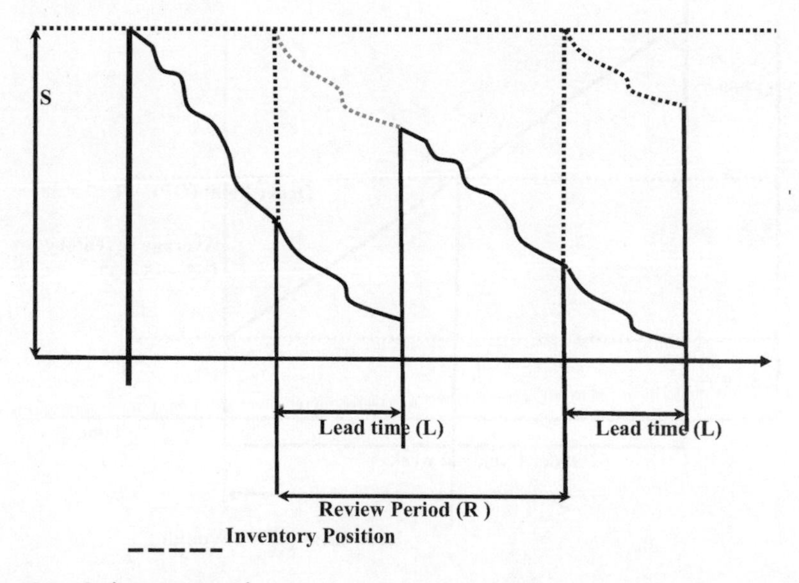

Figure 3.5 Order up to Level

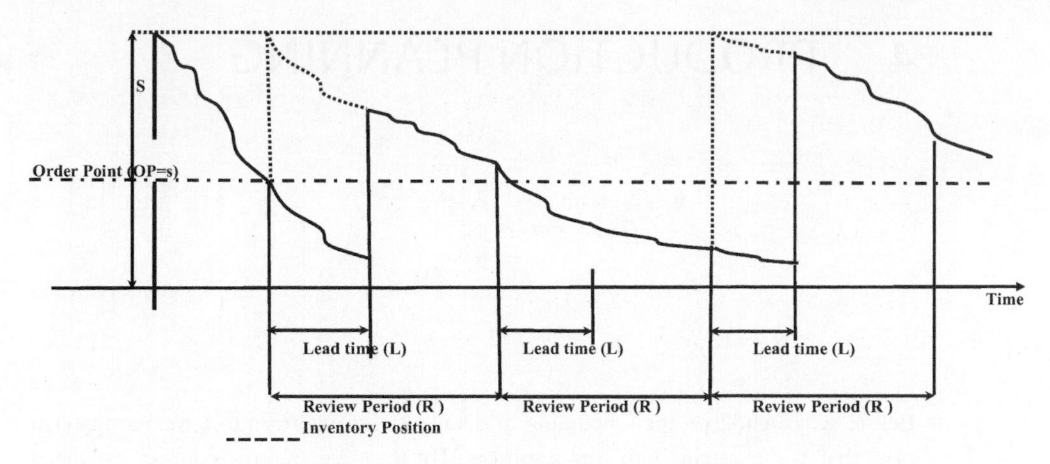

Figure 3.6 (R,s,S) Model

This has only been a short introduction to the management of inventory; there are whole long books written on the subject (Silver, 1998). However, it should be enough so that you can appreciate the role of inventory as we explore planning and scheduling in the following chapters.

4 PRODUCTION PLANNING

Before we can address the scheduling of detailed tasks, we need to have a long-term plan that is consistent with our resources. By resources in this context we mean those that require a relatively long time to acquire. Examples of these are operating rooms in hospitals, automated assembly equipment in factories, airplanes for airlines, and so forth. We can further segregate the time scale for these acquisitions. For the purposes of this text we will assume that large acquisitions such as those mentioned earlier are in place. Less time-consuming activities, such as hiring personnel or arranging for work to be done by a third party, can be included in our approach to planning. For the sake of convenience, we will break our planning into three phases—plans that are made a year in advance, those that plan for about three months, and the remaining ones that can be immediate or about up to two weeks.

This chapter concerns itself with the 1-year horizon. As most of the mathematics and practices in this area developed in the context of manufacturing, much of the terminology refers to it. However, the principles and methods are equally applicable to transportation, healthcare, distribution, mining, and any other area you can think of.

The planning process is driven by four forces:

1. Demand (orders) from customers and/or our own forecasts of requirements;
2. Our available resources;
3. Our ability to alter the near-term resources;
4. Any limitation imposed by practicality (such as work hours in a day) or management directives (such as limits on overtime).

Each of these may either be known or estimated to the best of our ability. Perhaps it is best to begin our study with an example.

Example 4.1 A manufacturing operation consists of the production of a single product. The demand, or orders for the product, is reasonably well known for the coming 12 months (our horizon). We know how many hours of labor are required to produce each unit of product. We also know how many workers are available at the beginning of the horizon. Further, the cost of carrying inventory from one month to the next and the costs of laying off or hiring a worker are all known. Our objective is to minimize the total cost of satisfying all of the demand by the end of the horizon (a year in this case).

To put this example into further perspective, let's examine the problem in general. For each month we need to decide how many units of the product to produce. At its simplest this would require producing exactly what the demand called for. There are three possibilities—the demand is either less, the same, or more than our ability to produce with our existing labor force. We have the following choices in case it is more:

1. Forgo satisfying the demand altogether;
2. Hire additional workers to meet the demand;
3. Convince the customer to accept a smaller quantity now, and accept the delivery of the balance in the next month (i.e., allow shortages);
4. Have our existing staff work extra hours (overtime);
5. Sublet (farm out, offload, subcontract) the shortage to a third party;
6. A combination of all of the above.

If the demand and our ability to produce are the same, we do not have to take any action. Finally, if the demand is less, the following options are open to us:

1. Let some of our staff go (lay off, fire, terminate, furlough);
2. Produce up to our ability and use the excess for future demand (inventory);
3. Produce more than our ability with the same actions available under the first case.

To determine which of these choices to exercise, we need to define a measurable objective. The most straightforward is the total cost of the annual plan. Naturally, we are subject to a number of restrictions (usually referred to as constraints), the most common of which are as follows:

1. The total demand for the product must be satisfied by the end of the planning horizon or the customer will not accept late shipments;
2. There are a limited number of hours available in each month (usually 160 per worker);
3. Each unit of product requires a specific number of labor hours to produce it;
4. The number of extra hours (overtime) that a worker can perform in a month is specified.

As we shall see later, each of the ideas mentioned so far may be modified according to the conditions of our specific problem. We are now in a position to specify our first example in detail:

Given: d_t—demand for the product in month t, $1 \le t \le 12$
C_s—Cost of subcontracting a unit of product
W_o—Initial number of workers before month 1
I_o—Inventory of the product before month 1
n—Hours required to produce a unit of product
M—The number of regular hours a worker can work in a month
C_I—The cost of carrying a unit of product from one month to the next

C_H—Cost of hiring a worker
C_F—Cost of laying off a worker
R—Cost of one regular hour of labor
O—Cost of one hour of overtime labor
O_{TL}—Number of overtime hours a worker may work in a month

But how are we to solve the problem? It is important to first precisely define the variables that are under our control:

X_t—The number of units of product to produce in each month (does not include subcontracted units)
I_t—The amount of inventory to carry from month t to t + 1
H_t—The number of workers to hire in and for month t
F_t—The number of existing workers to terminate at the beginning of month t
O_t—The total number of overtime hours worked in month t
S_t—The total number of units subcontracted in month t

Once we have assigned numbers to each of the given items (Figure 4.1) we need to decide just exactly how to minimize the total cost. Figure 4.2 shows the spreadsheet with all variables entered, as well as all conditions satisfied and an arbitrary (guessed) solution. We also express our previous definitions in equation form.

Intial Workforce	10	Cost per regular hour	$30
Initial Inventory	350	Cost per overtime hour	$45
Regular hours per worker per period	160	Cost of carrying inventory per period per unit	$12
Maximum Overtime hours per worker per period	30	Cost of subcontracting per unit	$80
Hours per product per worker	2	Cost of hiring a new worker	$1,000
		Cost of terminating a worker	$2,000

Figure 4.1 Data for the Planning Example 4.1

	D	E	F	G	H	I	J	K	L	M	N	O	P	Q	R
11									Period						
12			1	2	3	4	5	6	7	8	9	10	11	12	
13	Demand		900	905	848	627	766	828	696	807	857	713	661	965	9573
14															
15	Regular hours Production		550	892	848	627	766	828	696	807	857	713	661	887	
16	Subcontracted units		0	0	0	0	0	0	0	0	0	0	0	0	
17	Overtime hours per worker		0	2.4	0	0	0	0	0	0	0	0	0	13	
18	Total units available		900	905	848	627	766	828	696	807	857	713	661	965	9573
19	Number Hired		0	3	0	0	2	1	0	2	1	0	0	4	
20	Number terminated		2	0	0	3	0	0	1	1	0	3	0	1	
21	Inventory at beginning	350	0	0	0	0	0	0	0	0	0	0	0	0	
22															
23															
24	Work Force	10	8	11	11	8	10	11	10	11	12	9	9	12	
25	Work Balance		180	2.73	64	26	68	104	208	146	206	14	118	302	
26	Inventory Balance		0	0	0	0	0	0	0	0	0	0	0	0	
27	Overtime balance		210	327.6	330	240	300	330	300	330	360	270	270	347	
28	Demand Balance														0
29															
30	Cost of Hiring		$0	$3,000	$0	$0	$2,000	$1,000	$0	$2,000	$1,000	$0	$0	$4,000	
31	Cost of terminating		$4,000	$0	$0	$6,000	$0	$0	$2,000	$2,000	$0	$6,000	$0	$2,000	
32	Cost of inventory		$0	$0	$0	$0	$0	$0	$0	$0	$0	$0	$0	$0	
33	Cost of subcontracting		$0	$0	$0	$0	$0	$0	$0	$0	$0	$0	$0	$0	
34	Cost of regular hours		$38,400	$52,800	$52,800	$38,400	$48,000	$52,800	$48,000	$52,800	$57,600	$43,200	$43,200	$57,600	
35	Cost of Overtime hours		$0	$1,203	$0	$0	$0	$0	$0	$0	$0	$0	$0	$7,020	
36	Total Cost		$42,400	$57,003	$52,800	$44,400	$50,000	$53,800	$50,000	$56,800	$58,600	$49,200	$43,200	$70,620	$628,823

Figure 4.2 Manual Spreadsheet Solution for Satisfying the Demand Exactly in Each Month

The size of the work force (worker balance) $W_t = W_{t-1} + H_t - F_t$ for each t

The inventory balance $I_t = I_{t-1} + X_t - d_t + S_t$ for each t

The work balance, the hours required to produce the units must be less than or equal to the available hours $X_t \leq W_t (M + O_t) / n$ for each t

Demand satisfaction $\sum_{t=1}^{12} d_t = \sum_{t=1}^{12} (X_t + S_t)$

Overtime limit $O_t \leq O_{TL}$ for each t

All variables must be positive $X_t, I_t, H_t, F_t, S_t \geq 0$ for each t

The total cost = sum of the cost of (regular hours, overtime hours, hiring, firing, inventory, subcontracting)

$$\sum_{t=1}^{12} \left[RW_t + OO_t + H_t C_H + F_t C_F + I_t C_I + S_t C_S \right]$$

It is extremely unlikely that a guess will be optimal. In fact, completing the guesses is not that simple because we have to satisfy all the conditions. At this point you should try to better my guess by creating the spreadsheet and producing your own estimates without recourse to any mathematical techniques.

Solving for the Optimal Cost

Fortunately there is an excellent method available, as long as all our variables and conditions are linear, i.e., they do not contain powers or cross products. The method is referred to as linear programming (LP or the simplex method, which was developed by Dantzig in 1947). It relies on the principle that no matter what set of values for each variable we start with, the solution always proceeds to a better set of variables until it is unable to do so. And the stopping point is guaranteed to be optimal. Most of our problems, however, are not linear and optimality is not guaranteed, although usually achieved. Solutions generally depend on the starting values of the variables, so it is a good idea to start with a more or less feasible solution. Naturally, there is a large body of mathematics behind the method, which is beyond the content of this book. There are many references that have all these details. Suffice it to say that we have the simple expedient to use Excel's Solver add-in as long as the problem is small enough. Much more sophisticated software tools are commercially available to solve real applications. The appropriate dialog box for our example is shown in Figure 4.3. The resulting solution is shown in Figure 4.4. Because it is an integer problem for the work force changes and not the production numbers (answers were rounded), there is a possibility that slightly better solutions exist.

It is relatively straightforward to extend the basic problem to more than one product by adding the subscript i to the variables and constraints. Many operations periodically choose a method other than general optimization. There are two commonly used methods:

1. Chase—where each demand is exactly satisfied by varying the number of workers;
2. Level—where demand is satisfied by varying the inventory and the work force and keeping the production constant.

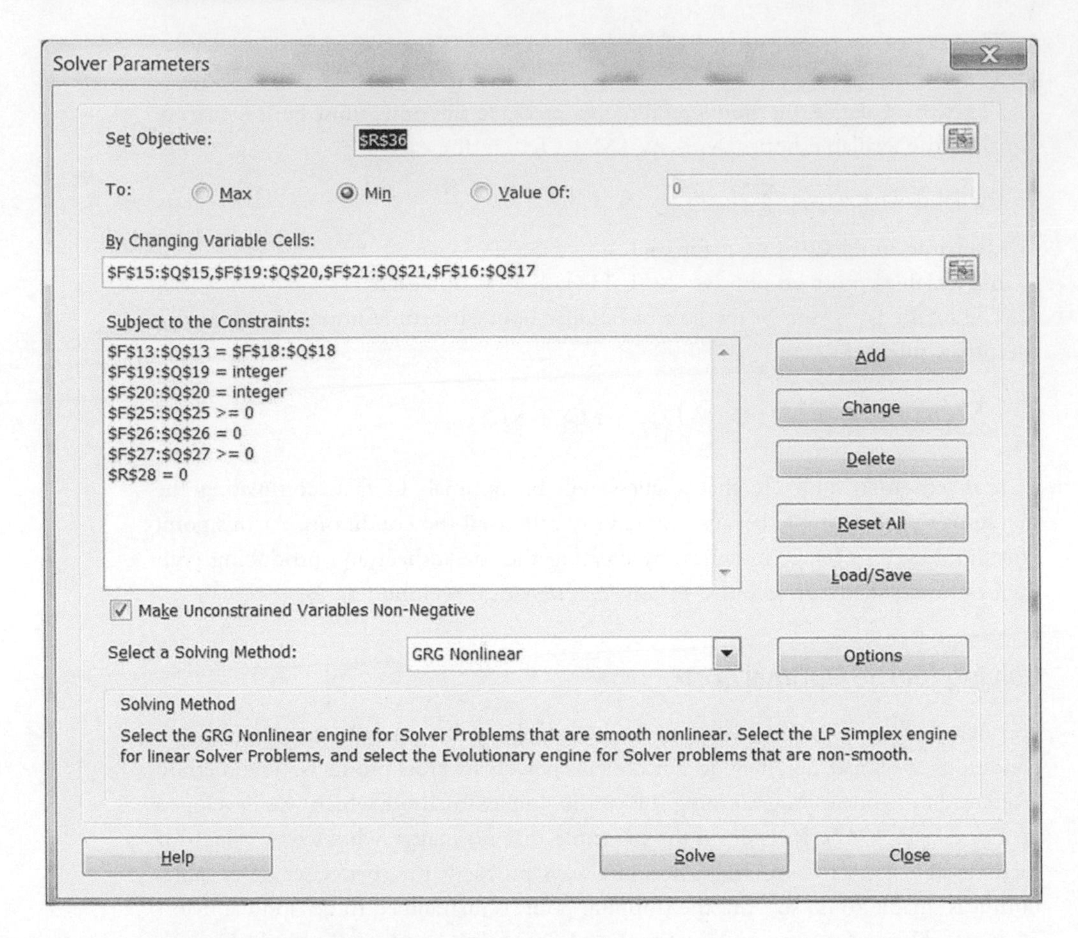

Figure 4.3 Excel Solver Dialog Box for Example 4.1

A chase method may also modify overtime and offloading ($X_{it} + S_{it} = d_{it}$ for each i and t). This has the benefit of creating zero inventory—a very desirable outcome. Unfortunately, the method is rarely successful in practice as controlling the variables and the available material is very difficult. The most common outcome is frequent late deliveries (shortages) for which our model does not yet account.

The level method produces the same number of units of each product in every period ($X_{it} + S_{it} = X_{it} + S_{it}$ for each i and t). The obvious benefit is no disruption to the routine production of the products and perhaps a constant work force. However, higher inventories and shortages tend to be introduced.

Example 4.2 Level approach to example 4.1. We have to rewrite some of the equations to make the production constant after the first period. In this instance we have also fixed the work force after the first period. Figures 4.5 and 4.6 have the dialog box and the resulting solution. This is an excellent example where

	D	E	F	G	H	I	J	K	L	M	N	O	P	Q	R
11									Period						
12			1	2	3	4	5	6	7	8	9	10	11	12	
13	Demand		900	905	848	627	766	828	696	807	857	713	661	965	9573
14															
15	Regular Production		550	892	848	627	766	828	696	807	855	713	661	889	
16	Subcontracted units		0	0	0	0	0	0	0	0	0	0	0	1	
17	Overtime hours per worker		0	2.3	0	0	0	0	0	0	0.45	0	0	14	
18	Total units available		900	905	848	627	766	828	696	807	857	713	661	965	9573
19	Number Hired		0	2	1	1	0	1	0	2	1	0	0	2	
20	Number terminated		1	0	1	2	0	0	2	0	1	1	1	0	
21	Inventory at beginning	350	0	0	0	0	0	0	0	0	0	0	0	0	
22															
23															
24	Work Force	10	9	11	11	10	10	11	9	11	11	10	9	11	
25	Work Balance		340	1	64	346	68	104	48	146	55	174	118	131	
26	Inventory Balance		0	0	0	0	0	0	0	0	0	0	0	0	
27	Overtime balance		270	327.7	330	300	300	330	270	330	329.549	300	270	316	
28	Demand Balance														0
29															
30	Cost of Hiring		$0	$2,000	$1,000	$1,000	$0	$1,000	$0	$2,000	$1,000	$0	$0	$2,000	
31	Cost of terminating		$2,000	$0	$2,000	$4,000	$0	$0	$4,000	$0	$2,000	$2,000	$2,000	$0	
32	Cost of inventory		$0	$0	$0	$0	$0	$0	$0	$0	$0	$0	$0	$0	
33	Cost of subcontracting		$0	$0	$0	$0	$0	$0	$0	$0	$0	$0	$0	$73	
34	Cost of regular hours		$43,200	$52,800	$52,800	$48,000	$48,000	$52,800	$43,200	$52,800	$52,800	$48,000	$43,200	$52,800	
35	Cost of Overtime hours		$0	$1,125	$0	$0	$0	$0	$0	$0	$223	$0	$0	$6,705	
36	Total Cost		$45,200	$55,925	$55,800	$53,000	$48,000	$53,800	$47,200	$54,800	$56,023	$50,000	$45,200	$61,579	$626,527

Figure 4.4 Solver Solution for Satisfying the Demand Exactly in Each Month for Example 4.1

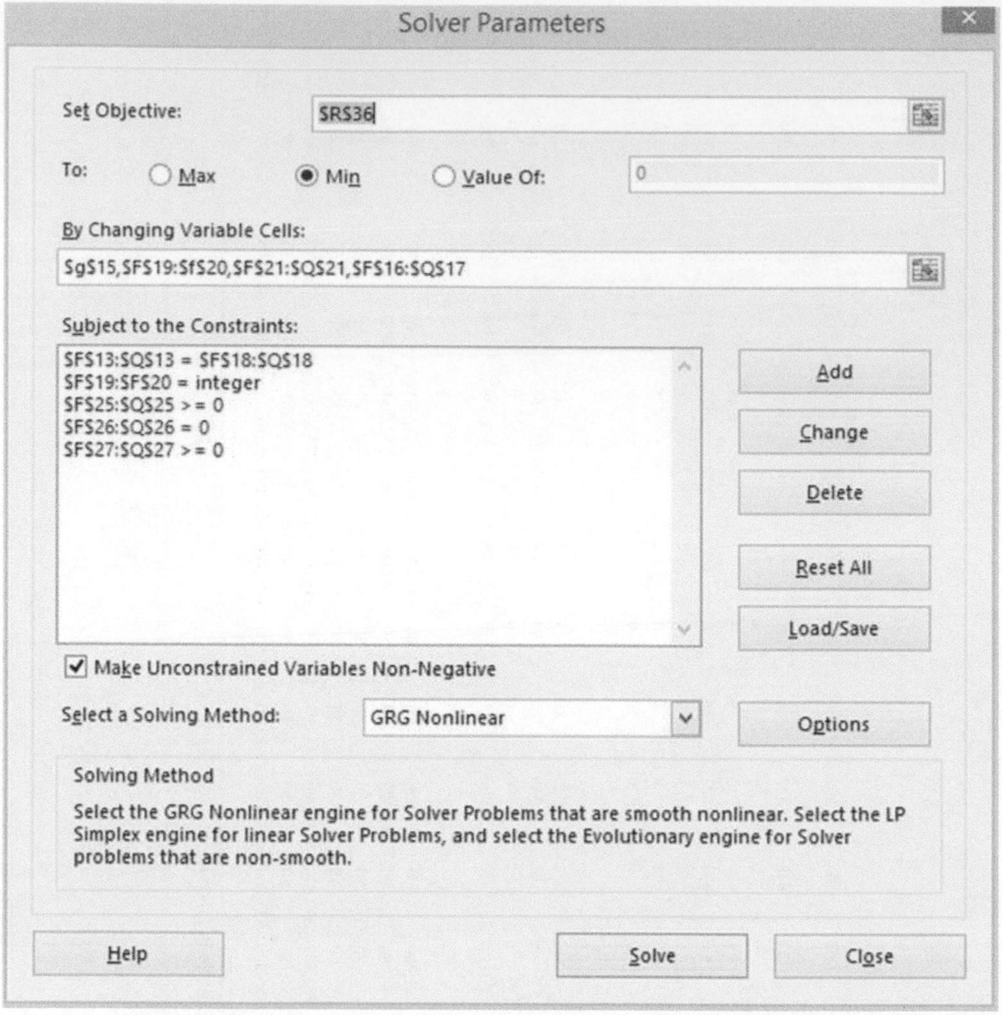

Solver Parameters

Set Objective: $R$36

To: ○ Max ◉ Min ○ Value Of: 0

By Changing Variable Cells:

g15,$F$19:f20,$F$21:$Q$21,$F$16:$Q$17

Subject to the Constraints:

$F$13:$Q$13 = $F$18:$Q$18
$F$19:$F$20 = integer
$F$25:$Q$25 >= 0
$F$26:$Q$26 = 0
$F$27:$Q$27 >= 0

Add

Change

Delete

Reset All

Load/Save

☑ Make Unconstrained Variables Non-Negative

Select a Solving Method: GRG Nonlinear

Options

Solving Method

Select the GRG Nonlinear engine for Solver Problems that are smooth nonlinear. Select the LP Simplex engine for linear Solver Problems, and select the Evolutionary engine for Solver problems that are non-smooth.

Help Solve Close

Figure 4.5 Dialog Box for the Level Solution (Example 4.2)

judgment is needed in interpreting the solution. In this instance, because of the relative value of subcontracting and terminating employees and the cost of an hour of labor, a large quantity of product is to be subcontracted. This would result in a substantial increase in per hour labor costs (see Appendix A) and not be in our best interest.

This leaves us with one important consideration—how do we treat the cost of not satisfying a demand in the month requested. Usually, there is no immediate financial penalty—we tell the customer that a shipment will be delayed and the customer grudgingly accepts the condition because he has already built in a safety factor (see the chapter on inventory and consider it from the customer's point of view) and because more often than not it is difficult or impossible to obtain the

	D	E	F	G	H	I	J	K	L	M	N	O	P	Q	R
11									Period						
12			1	2	3	4	5	6	7	8	9	10	11	12	
13	Demand		900	905	348	627	766	828	696	807	857	713	661	965	9573
14															
15	Regular Production		550	627	627	627	627	627	627	627	627	627	627	627	
16	Subcontracted units		0	257	221	0	139	201	69	180	228	86	34	296	
17	Overtime hours per worker		0	5.1	0	0	0	0	0	0	0.38	0	0	10	
18	Total units available		900	905	848	627	766	828	696	807	857	713	661	965	9573
19	Number Hired		0												
20	Number terminated		2												
21	Inventory at beginning	350	0	0	0	0	0	0	0	0	0	0	0	0	
22															
23															
24	Work Force	10	8	8	8	8	8	8	8	8	8	8	8	8	
25	Work Balance		180	67	26	26	26	26	26	26	29	26	26	109	
26	Inventory Balance		0	0	0	0	0	0	0	0	0	0	0	0	
27	Overtime balance		240	234.9	240	240	240	240	240	240	240	240	240	230	
28	Demand Balance														0.00
29															
30	Cost of Hiring		$0	$0	$0	$0	$0	$0	$0	$0	$0	$0	$0	$0	
31	Cost of terminating		$4,000	$0	$0	$0	$0	$0	$0	$0	$0	$0	$0	$0	
32	Cost of inventory		$0	$0	$0	$0	$0	$0	$0	$0	$0	$0	$0	$0	
33	Cost of subcontracting		$0	$20,560	$17,680	$0	$11,120	$16,080	$5,520	$14,400	$18,240	$6,880	$2,720	$23,680	
34	Cost of regular hours		$38,400	$38,400	$38,400	$38,400	$38,400	$38,400	$38,400	$38,400	$38,400	$38,400	$38,400	$38,400	
35	Cost of Overtime hours		$0	$1,845	$0	$0	$0	$0	$0	$0	$135	$0	$0	$3,735	
36	Total Cost		$42,400	$60,805	$56,080	$38,400	$49,520	$54,480	$43,920	$52,800	$56,775	$45,280	$41,120	$65,815	$607,395

Figure 4.6 Solver Solution for the Level Work Force (Example 4.2)

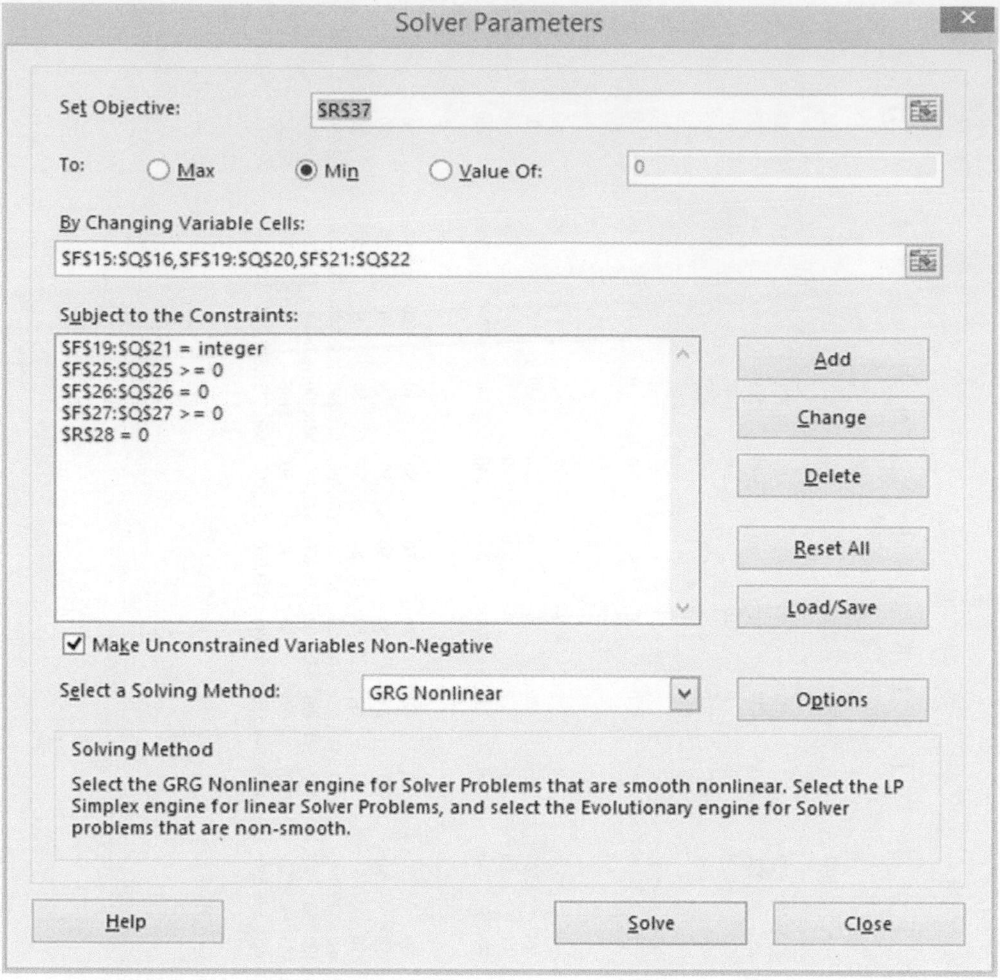

Solver Parameters

Set Objective: $R$37

To: ○ Max ◉ Min ○ Value Of: 0

By Changing Variable Cells:

$F$15:$Q$16,$F$19:$Q$20,$F$21:$Q$22

Subject to the Constraints:

$F$19:$Q$21 = integer
$F$25:$Q$25 >= 0
$F$26:$Q$26 = 0
$F$27:$Q$27 >= 0
$R$28 = 0

Add

Change

Delete

Reset All

Load/Save

☑ Make Unconstrained Variables Non-Negative

Select a Solving Method: GRG Nonlinear

Options

Solving Method

Select the GRG Nonlinear engine for Solver Problems that are smooth nonlinear. Select the LP Simplex engine for linear Solver Problems, and select the Evolutionary engine for Solver problems that are non-smooth.

Help Solve Close

Figure 4.7 Dialog Box for Example 4.3

same product from someone else. However, frequent such occurrences will eventually lose the customer and/or damage our reputation with other customers. Therefore it is important to assume some penalty that will minimize such occurrences.

Example 4.3 We will redo our original scenario while allowing shortages and optimize while not insisting or either chase or level. For this we do need one more variable for shortages, SG_{it}, and the monthly shortages are added as a variable and the inventory balance equation adjusted for it. The shortage cost is assumed to be $200/unit.

The inventory balance $I_t - SG_t = I_{t-1} + X_t - d_t + S_t - SG_{t-1}$ for each t.

The dialog box and the solution are shown in Figures 4.7 and 4.8.

The conditions of each operation would dictate whether one chooses level, chase, or optimization, but if practicable, one should choose optimization.

	D	E	F	G	H	I	J	K	L	M	N	O	P	Q	R
11									Period						
12			1	2	3	4	5	6	7	8	9	10	11	12	
13	Demand		900	905	848	627	766	828	696	807	857	713	661	965	9573
14															
15	Regular Production		800	800	640	640	720	720	720	800	800	800	800	800	
16	Subcontracted units		0	0	0	0	0	134	3	37	2	2	2	4	
17	Overtime hours per worker		0	0	0	0	0	0	0	0	0	0	0	0	
18	Total units available		1150	800	640	640	720	854.035	722.864	837.288	802	801.6	801.6	804	9573
19	Number Hired		0	0	0	0	1	0	0	1	0	0	0	0	
20	Number terminated		0	0	2	0	0	0	0	0	0	0	0	0	
21	Inventory at beginning	350	250	145	0	13	0	0	24	17	0	87	226	61	
22	Shortage at beginning		0	0	63	0	33	108	0	0	40	0	0	0	
23															
24	Work Force	10	10	10	8	8	9	9	9	10	10	10	10	10	
25	Work Balance		0	0	0	0	0	0	0	0	0	0	0	0	
26	Inventory Balance		0	0	0	0	0	0	0	0	0	0	0	0	
27	Overtime balance		300	300.0	240	240	270	270	270	300	300	300	300	300	
28	Demand Balance														0.00
29															
30	Cost of Hiring		$0	$0	$0	$0	$1,000	$0	$0	$1,000	$0	$0	$0	$0	
31	Cost of terminating		$0	$0	$4,000	$0	$0	$0	$0	$0	$0	$0	$0	$0	
32	Cost of inventory		$3,000	$1,740	$0	$156	$0	$0	$288	$204	$0	$1,044	$2,712	$732	
33	Cost of subcontracting		$0	$0	$0	$0	$0	$10,723	$229	$2,983	$128	$128	$128	$321	
34	Cost of regular hours		$48,000	$48,000	$38,400	$38,400	$43,200	$43,200	$43,200	$48,000	$48,000	$48,000	$48,000	$48,000	
35	Cost of shortage		$0	$0	$12,600	$0	$6,600	$21,600	$0	$0	$8,000	$0	$0	$0	
36	Cost of Overtime hours		$0	$0	$0	$0	$0	$0	$0	$0	$0	$0	$0	$0	
37	Total Cost		$51,000	$49,740	$55,000	$38,556	$50,800	$75,523	$43,717	$52,187	$56,128	$49,172	$50,840	$49,053	$621,716

Figure 4.8 Solver Solution for the Case of Shortages (Example 4.3)

5 MANUFACTURING REQUIREMENTS PLANNING

Here we consider a time span of about three months. The assumption is that most materials can be acquired within 3 months. The planning period certainly needs to be long enough to encompass the total time required to produce the item with the longest lead time from ordering the material to completing the product. Our objective is to produce a sufficiently detailed plan that will provide the necessary materials and resources to deliver specific products for which customers have asked, or to produce the specific products on which we have decided based on our forecast (The subject of forecasting is beyond the scope of this text—many good texts are available).

To do this we will require a substantial set of data:

1. Customer orders;
2. Our own planned production for each item;
3. Available resources of personnel and machinery;
4. Resources required to produce each item, including the routing or process plan for each item and the time required to perform each task that contributes to the final product;
5. Detailed bills of materials for each item;
6. A complete list of all required parts and the lead time for each of these.

Most operations depend on commercial software that processes these items and is generally referred to as Enterprise Resource Planning, or ERP.

This software performs many functions, among which is Manufacturing Resource Planning (MRP), which is a forerunner of ERP and is of interest to us here. Because the remainder of this book concerns itself with scheduling that is a follow-on and supplement to the output of the MRP, we need to have an understanding of how MRP does this. (We only consider features that have a direct bearing on scheduling—many references are available that describe all the features of MRP).

In Chapter 4 we learned how a company would decide how many of each product to build in each period, usually a month, based on estimates of demand. We did not address how this would actually be accomplished. The process of going from the annual plan to a detailed plan for each product, usually called the master schedule, is accomplished by the MRP system.

The function of MRP is to take the customer orders and our forecasts and convert them to a production plan that specifies when the production of each part is to start and provides an order date for all required materials so that they arrive in time for the planned starting time of each phase of the product.

The MRP Record

The software starts by creating what is called an MRP record for each independent item. By an independent item we mean one that is delivered to a customer. A car is an example of an independent item, while all the parts that make up the car, such as wheels, are dependent items. Sometimes parts become independent items when they are supplied as spares for repairs. The number of any dependent item that is required is a function of the number of cars.

The function of the MRP record is to show the demand for the item in each period (usually weeks), any forecasts for the item, the amount of inventory on hand, when all the required parts have to arrive and how many of the item are to be delivered against the demand, as well as how many of the item may be promised toward new orders without changing the current plan. There are two significant time periods in an MRP record. The first is the lead time, that is the time required to produce the item from the required components. The second is the demand time fence. This defines the time before which we only consider demand and ignore the forecast. After the time fence we take the larger of the demand and forecast to calculate our inventory and required completions of the item. We do this because we assume that inside the demand time fence we have received all the orders we could reasonably expect, but after the time fence we do expect more demand to get us closer to the forecast. While available to promise (ATP) is calculated in many different ways in different companies, we will restrict ourselves to just one method. In this method, we only look back to prior periods in the first period and assume that the inventory before the first period is available to fulfill demands up to the next planned completion of the item. Anything that is left after satisfying that demand becomes available to promise. We never consider forecast in calculating the ATP. In any subsequent period in which there is a completion of the item, we calculate the ATP by subtracting all demand until the next completion from the number completed. I have used the term "completed" to be clear that it is readiness of the item to be delivered to the customer, but in MRP language these are referred to as planned receipts and we will do so from here on. Planned order refers to the time the required components for the item are to be received. An example that encompasses these features is shown in Figure 5.1. The timing of the receipts is determined by not letting the projected on hand go below zero. The quantity used for receipts in this case is one that would have been calculated with the methods described in Chapter 3 for more or less constant demand. However, other methods are used when demand is not constant, which we will cover a bit later in the chapter.

But first we want to complete the functions of the MRP record. There also has to be an MRP record for each dependent item. This record does not have a forecast,

		Period											
		1	2	3	4	5	6	7	8	9	10	11	12
Demand		100	110	95	88	108	75	70	60	45	0	0	0
Forecast		90	100	100	100	100	110	115	120	110	100	100	100
Projected on Hand (POH)	110	10	200	105	17	209	99	284	164	54	254	154	54
Planned receipt			300			300		300			300		
Planned Order			300		300			300					
Available to promise		10	7			117		125			300		

Demand time fence (between periods 4 and 5)

The lead time is 3 periods. The receipt in period 2 must have been ordered before period 1
Note that the ATP in period 1 is 110-100, in period 2 it is 300-110-95-88

Figure 5.1 An Example of an MRP Record for an Independent Item

		Period											
		1	2	3	4	5	6	7	8	9	10	11	12
Demand		0	900	0	900	0	0	900	0	0	0	0	0
Projected on Hand (POH)	850	850	1450	1450	550	550	550	1150	1150	1150	1150	1150	1150
Planned receipt			1500					1500					
Planned Order						1500							

The lead time is 2 periods. The receipt in period 2 was planned before period 1

Figure 5.2 An Example of an MRP Record for an Item Depending on the Item in Figure 5.1

a demand time fence, or an available to promise—all these are driven by the independent item. The demand for a dependent item is tied to the record of the independent by the planned order of the independent item. Figure 5.2 shows an example that is tied to the record in Figure 5.1 and is based on the fact that three of this dependent item are required for each independent one.

The procedures for determining the quantities to be used in the MRP record are referred to as lot sizing. The most straightforward of these is lot for lot (LOL), that is, the quantity is the same as the demand, just as we learned in planning when we referred to it as the chase plan. There are a number of others, such as the fixed quantity we showed in the last example. This is a fairly good method as long as the demand is more or less constant and we can use the EOQ equation from Chapter 3. One convenient method for uneven demand is least unit cost (LUC). For other methods refer to Silver (1998). This is obtained by averaging the ordering and inventory cost while accumulating requirements until the unit cost increases. This is best shown by an example. Let's assume that the ordering cost is \$35 and the inventory carrying cost is \$0.15 per unit per period and the demand is given in Figure 5.3. As an example the unit cost in period 7 if we were to order for four periods is calculated by adding the ordering cost to the inventory cost of carrying the demand in period 8 for one period, the inventory carrying cost in period 9 for two periods, and the inventory cost for carrying the demand in period 10 for three periods. The total cost for this approach is six orders at \$35 plus the inventory carrying cost of the inventory in the row labeled inventory (\$0.15*602) for a total of \$300.30. Contrast this with the cost in the example of Figure 5.1, which is \$380.60.

	Period											
	1	2	3	4	5	6	7	8	9	10	11	12
Demand	100	85	120	55	70	104	65	88	60	101	150	76
Inventory	85	0	55	0	104	0	148	60	0	150	0	0
Order	185		175		174		213			251		76
Order for one period												
Quantity	100		120		70		65			101		
Cost of ordering	$35		$35		$35		$35			$35		
Cost of Inventory	$0		$0		$0		$0			$0		
Unit cost	$0.350		$0.292		$0.500		$0.538			$0.347		
Order for two periods												
Quantity	185		175		174		153			251		
Cost of ordering	$35		$35		$35		$35			$35		
Cost of Inventory	$13		$8		$16		$13			$23		
Unit cost	**$0.258**		**$0.247**		**$0.291**		$0.315			**$0.229**		
Order for three periods	305		245		239		213			327		
Cost of ordering	$35		$35		$35		$35			$35		
Cost of Inventory	$49		$29		$35		$31			$45		
Unit cost	$0.275		$0.262		$0.293		**$0.311**			$0.246		
Order for four periods							314					
Cost of ordering							$35					
Cost of Inventory							$76.65					
Unit cost							$0.356					

Figure 5.3 Demand and Calculations for the Least Unit Cost Example

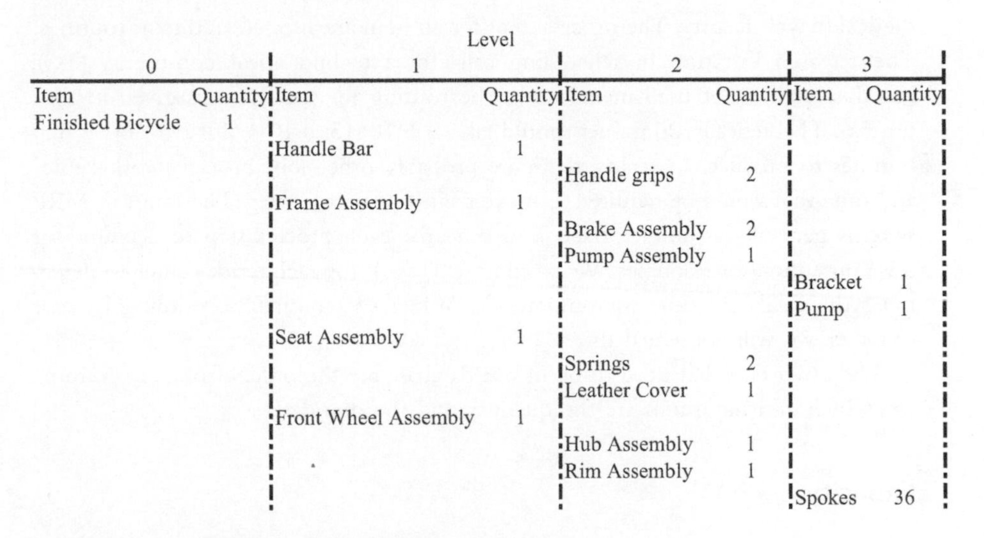

Level 0 Item	Quantity	Level 1 Item	Quantity	Level 2 Item	Quantity	Level 3 Item	Quantity
Finished Bicycle	1						
		Handle Bar	1				
				Handle grips	2		
		Frame Assembly	1				
				Brake Assembly	2		
				Pump Assembly	1		
						Bracket	1
						Pump	1
		Seat Assembly	1				
				Springs	2		
				Leather Cover	1		
		Front Wheel Assembly	1				
				Hub Assembly	1		
				Rim Assembly	1		
						Spokes	36

Figure 5.4 Simplified Partial Bill of Materials for a Bicycle

The Database Structure for MRP

As you can surmise from the relationship between two items in Figures 5.1 and 5.2, we need to know what constitutes each and every assembly and subassembly for all products. This structure is the bill of materials (BOM). A much simplified partial example for a bicycle is shown in Figure 5.4.

Once we have bills of materials for all our products, we need to define the process steps that each item has to go through and the associated work centers

WorkCenter	Name
A	Plastic Molding
B	Metal cutting
C	NC Machining
D	Welding
E	Plating
F	Subassembly
G	Final Assembly

Figure 5.5 Typical Work Centers

Work Center	Step No.	Time per unit (minutes)	Set up Time (minutes)
B	1	15	5
D	2	20	10
E	3	0	60

Figure 5.6 Routing for the Bicycle Frame

where the processes are performed. Figure 5.5 lists the work centers for our hypothetical bicycle factory. The process steps for an item are also referred to as routings. The research literature in scheduling calls these technological constraints (TC), but this term is not used in industry. The routing for the frame is given in Figure 5.6. Theoretically, 20 frames should take 5 + 10*15 + 10 + 20*20 + 60 = 625 minutes to produce. In reality, there are probably other jobs in each work center and time that would be required to move from center to center. The input to MRP systems generally assumes a fixed lead time for each process step to account for this contention for resources. We would use 1 week for each work center to determine the total lead time for our bicycle. When we consider scheduling in later chapters, we will use actual times.

The other data that need to be in our database are the orders from the customers, which at a minimum are the quantity and the due date.

Capacity

A basic feature of MRP systems is that they assume that we have all the capacity in our resources that we need to accomplish our master schedule. There is a very good reason for this and the aforementioned fixed lead time. Currently, and probably for the foreseeable future, economically viable computers cannot produce a plan that both accounts for real-time scheduling and capacity utilization and minimization of costs in a reasonable time. There are current add-ons, called advanced production systems (APS), that can do this for parts of master schedules. We will discuss the functioning of these in later chapters. For the time being, we assume fixed lead times and check whether we have or do not have adequate capacity for each of our resources. If we do not, the system cannot help us—we have to decide what action to take. Actions in general include overtime, subcontracting, working extra shifts, or renegotiating due dates with the customers.

The given Master Production Schedule for products X and Y for 5 periods

Week	1	2	3	4	5
X	12	10	17	12	17
Y	23	25	25	18	22

Routing Information

Product	Work Center	Operation	Setup Hours	Run Time / Unit
X	100	1 of 1	0.5	0.425
Y	100	1 of 1	0.5	1.103
A	200	1 of 1	2	0.655
B	300	1 of 1	1	0.305
C	200	1 of 2	1	0.802
C	300	2 of 2	0.7	0.956

Bills of Material

X
A B

Y
A 2C

Capacity Requirements

Week	1	2	3	4	5
WC 100	31.47	32.83	35.80	25.95	32.49
WC 200	62.82	66.03	70.61	51.52	63.83
WC 300	49.34	52.55	54.69	39.78	48.95

Figure 5.7 Rough Cut Capacity Example

There are numerous methods for assessing capacity. We will cover two here and a third in more detail in the chapters on scheduling. The first is called rough cut capacity (RCC). It calculates the total hours required in a work center for all the products in a period, such as a month and compares it to the available hours. This ignores the timing of the tasks and consequently can be quite misleading, but in many cases can be adequate, especially if work tends to be evenly distributed over the periods. Figure 5.7 has an example of a rough cut capacity analysis involving two products and three work centers and five periods. The demand is due at the end of each period. For example, the usage in work center 200 in period 1 is calculated by: (demand for X plus demand for Y)(hours for A in work center 200) plus set up for A in 200 plus set up for C in 200 plus 2(demand for Y)(hours for C in 200) = (12 + 23)(0.655) + 2 + 1 + 2(23)(0.802) = 62.817. This of course ignores the fact that A and B and the two stages of C have to be performed before either X or Y can be made. That is why it is called rough cut.

The second method is capacity requirements planning (CRP), although it is an assessment, not a plan. It usually assigns a fixed period in which each task is to be accomplished and thus takes into account the levels in the bills of materials. Thus in the previous example the requirements for X and Y in period 5 are performed in work center 100 in period 5, the associated requirements for A, B, and the second step of C fall in period 4, and the first step for C in period 3. Consequently there are requirements for work centers 200 and 300 in periods before period 1. The

Capacity Requirements

Week	-1	0	1	2	3	4	5
WC 100			31.47	32.83	35.80	25.95	32.49
WC 200	37.89	66.03	66.03	59.38	57.94	27.55	
WC 300		49.34	52.55	54.69	39.78	48.95	

Figure 5.8 Capacity Requirements Planning Example

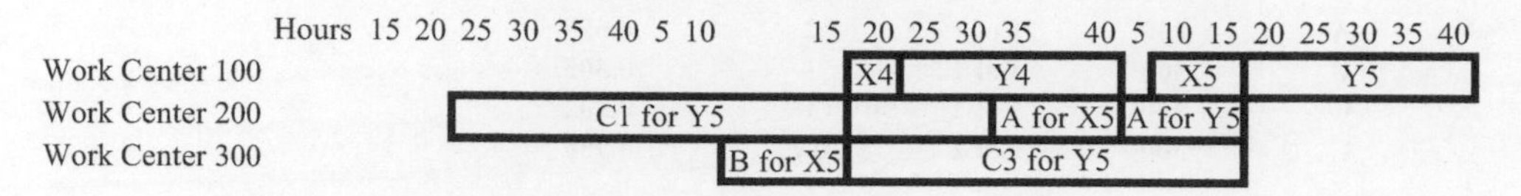

Figure 5.9 Finite Schedule for Period 5 of the Example

complete calculations for this method are shown in Figure 5.8. While the total requirement in each work center is the same as it was with rough cut capacity, the distribution is quite different.

The third method is to simulate the processes in sequence using real time. This is also referred to as finite scheduling. For this example we will assume the work centers 100 and 300 have a capacity of 40 hours per period, while work center 200 has 80 hours. In addition we assume that a work center can only work on one product at a time. In a real situation there would be other products using the same work centers, that is why the capacity requirements method allows for a whole period for each step of the process. A partial finite schedule is shown in Figure 5.9. For simplicity the hours have been rounded to the nearest 5 hours and only the full requirements for period 5 are shown. Notice that this presents a very different picture of the utilization of the work centers and would actually happen if there were no other products, there were no breakdowns, and we wanted the demand in period 5 to be delivered on time. This is a very strong argument for performing finite scheduling at least for a portion of our schedules. We now have sufficient background to begin our study of scheduling, which we left after the introductory chapter.

6 SCHEDULING PROBLEMS

The General Job Shop Scheduling Problem

If the theory of scheduling were simply concerned with the efficient reading of newspapers, no one would study it. I began with that example so that you might meet and attempt to solve a scheduling problem unhindered by the definitions and notations that are usually required and are introduced in this chapter. The terminology of scheduling theory arose in the processing and manufacturing industries. Therefore we will be talking about jobs and machines, even though in some cases the objects referred to bear little relation to either jobs or machines. For instance, in the example of the last section we will see that Albert, Bertrand, Charles, and Daniel are 'jobs,' while the newspapers are 'machines.' We begin by defining the general job shop problem. We will show that its structure fits many scheduling problems arising in business, computing, government, and the social services as well as those in industry.

Suppose that we have *n* jobs $\{J., J_2, \ldots J_n\}$ to be processed through m machines $\{M_1, M_2 \ldots M_m\}$. We can also refer to machines as processors. Each job must pass through each machine once and only once. The processing of a job on a machine is called an operation. The operation on the ith job by the jth machine is denoted by $\dot{o}_{ij}$. Technological constraints demand that each job should be processed through the machines in a particular order. For general job shop problems there are no restrictions upon the form of the technological constraints. Each job has its own processing order and this may bear no relation to the processing order of any other job. An important special case arises when all the jobs share the same processing order. In such circumstances we say that we have a flow shop problem (because the jobs flow between the machines in the same order). This distinction between job shops and flow shops will be made clear by the following examples.

Each operation o_{ij} takes a certain length of time to perform called the processing time and is denoted by p_{ij}. We include in p_{ij} any time required to adjust, or set up, the machine to process this job. We also include any time required to transport the job to the machine. We assume that the p_{ij} are fixed and known in advance. This brings us to an important restriction, which we make throughout most of this book. We assume that every numeric quantity is deterministic and known.

We also assume that the machines are always available, but do not necessarily assume the same for jobs. Some jobs may not become available until after the scheduling period has started. This is denoted by r_i, the ready time or release date of the ith job, i.e., the time at which J_i becomes available for processing but is not necessarily started. The general problem is to find a sequence, in which the jobs pass between the machines, that is:

1. Compatible with the technological constraints, i.e., a feasible schedule;
2. Optimal with respect to some criterion of performance.

Industrial Examples

Any manufacturing firm not engaged in mass production of a single item will have scheduling problems at least similar to that of the job shop. Each product will have its own route through the various work areas and machines of the factory. In the clothing industry, different styles have different requirements in cutting, sewing, pressing and packing. In steel mills, each size of rod or girder passes through the set of rollers in its own particular order and with its own particular temperature and pressure settings. In the printing industry, the time spent in typesetting a book will depend on its length, number of illustrations, and so on; the time spent in the actual printing will depend on both its length and the number printed; the time spent on binding will depend on the number printed; and the time spent in packaging will depend both on the number printed and the book's size. Thus a printer who must schedule the production of various books through his typesetting, printing, binding, and packaging departments faces a four-machine flow shop problem; for each department there is a machine and the jobs—i.e., the books—flow from typesetting to printing to binding to packaging.

The objectives in scheduling will vary from firm to firm and often from day to day. Perhaps the aim would be to maintain an equal level of activity in all departments so that expensive skills and machines were seldom idle. Perhaps it would be to achieve certain contractual target dates or simply to finish all the work as soon as possible. These and some other objectives will be discussed in detail later.

Albert and Friends

This is a four-job, four-machine problem. The jobs—Albert, Bertrand, Charles and Daniel—must be scheduled through the four machines—the *Financial Times*, the *Los Angeles Times*, the *Enquirer*, and the *New York Times*—in order to minimize the completion time at which the processing, here reading, is finished. There are ready times, namely, the times at which Albert and the others get up. Note that in this example the technological constraints do not demand that the roommates read the papers in the same order. Because of this we have a job shop problem. Had the constraints demanded that each read the papers in the same order, e.g., *Financial Times*, the *Los Angeles Times*, the *Enquirer*, and the *New York Times*, we would have had a flow shop problem.

Aircraft Queuing Up to Land

This is an *n*-job, one-machine problem, if we assume, as we do, that the number of aircraft arriving in a day is known. The aircraft are the jobs and the runway is the machine. Each aircraft has a ready time, namely the earliest time at which it can get to the airport's airspace and be ready to land. The objective here might be to minimize the average waiting time of an aircraft before it can land. Perhaps this average should be weighted by the number of passengers on each plane. Obviously in real life this problem models only part of the air traffic controllers' difficulties. Planes must take off, too. Also it ignores the uncertainty inherent in the situation. Aircraft suffer unpredictable delays and, moreover, the time taken to land, i.e., the processing time of each aircraft, will depend on the weather. Some airports have multiple runways, in which case we have a parallel-machine problem that will be treated in Chapter 13. We should also mention that in recent years the methods of predicting the arrival time of aircraft have been improved substantially.

Treatment of Patients in a Hospital

Again we ignore all randomness. Suppose we have *n* patients who must be treated by a surgeon. Then each must be

M_1—seen in the outpatient department;
M_2—received in the surgical department, prepared for the operation, and so forth;
M_3—be operated on;
M_4—recover, we hope, in intensive care.

Thus each patient (job) must be processed through each of four 'machines'—M_1, M_2, M_3, and M_4. It is perhaps confusing that only one of the operations of processing a patient through a 'machine' is called a surgical operation, but in scheduling theory all are operations. Because each patient must clearly 'flow' through the 'machines' in the order M_1, M_2, M_3, M_4, this is a flow shop problem. The objective here might be stated as: treat all patients in the shortest possible time, while giving priority to the most ill.

Other Scheduling Problems

Given these examples you should have no difficulty in fitting other practical problems into the general job shop structure. The following all fall into this pattern:

1. The scheduling of different programs on a computer;
2. The processing of different batches of crude oil at a refinery;
3. The repair of cars in a repair shop;
4. The manufacture of paints of different colors.

Assumptions

For the major part of this book we are making a number of assumptions about the structure of our scheduling problems. Some were mentioned explicitly earlier; others were implicit. Here we list all the assumptions.

1. *Each job is an entity.* Although each job is composed of distinct operations, no two operations of the same job may be processed simultaneously. We exclude certain practical problems, e.g., those in which components are manufactured simultaneously prior to assembly into the finished product.
2. *No pre-emption.* Once started, each operation must be completed before another operation may be started on that machine.
3. *Each job has* m *distinct operations, one on each machine.* We do not allow for the possibility that a job might require processing twice on the same machine. We insist that each job is processed on every machine; it may not skip one or more machines. Note that this latter constraint is *not* illusory. Although we could say that a job that skips a machine is processed upon that machine for zero time, we would still have a problem: where in the job's processing sequence this null operation should be placed. Because we do not allow pre-emption, the job could be delayed waiting for a machine that is not in fact needed.
4. *No cancellation.* Each job must be processed to completion.
5. *The processing times are independent of the schedule.* First, each set-up time is sequence independent, i.e., the time taken to adjust a machine for a job is independent of the job last processed. Second, the times to move jobs between machines are negligible.
6. *In-process inventory is allowed*, i.e., jobs may wait for their next machine to be free. This is not a trivial assumption. In some problems processing of jobs must be continuous from operation to operation. In steel mills one literally has to strike while the iron is hot.
7. *There is only one of each type of machine.* We do not allow that there might be a choice of machines in the processing of a job. This assumption eliminates, among others, the case where certain machines have been duplicated to avoid bottlenecks.
8. *Machines may be idle.*
9. *No machine may process more than one operation at a time.*
10. *Machines never break down and are available throughout the scheduling period.*
11. *The technological constraints are known in advance and are unchangeable.*
12. *There is no randomness.* In particular:
    a. The number of jobs is known and fixed;
    b. The number of machines is known and fixed;
    c. The processing times are known and fixed;
    d. The ready times are known and fixed;
    e. All other quantities needed to define a particular problem are known and fixed.

Occasionally we will relax one or two of these assumptions in specific examples and state so explicitly.

Performance Measures

It is not easy to state our objectives in scheduling. They are numerous, complex, and often conflicting. In addition, the mathematics of our problem can be extremely difficult with even the simplest of objectives. So we do not try to be exhaustive. Instead we indicate in general terms a few of the criteria by which we might judge our success. These criteria will be sufficient to motivate the mathematical definitions of the performance measures that we use.

Obviously we should prefer to keep promised delivery dates. Otherwise good will would surely be lost and there might be financial penalties as well. We should also try to minimize the overall length of the scheduling period, because once all the jobs have been completed the machines may be released for other tasks. We should try to minimize the time for which the machines are idle; idle machines mean idle capital investment. We should try to minimize inventory costs, and by these we do not mean just the cost of storing the finished product. There are also the costs of storing the raw materials and any partially processed jobs that must wait between machines. We might try to ensure a uniform rate of activity throughout the scheduling period so that demands for labor and power are stable. Conversely, it might be desirable to concentrate the activity into periods when either labor or power is particularly cheap.

Before we can define performance measures in precise mathematical terms, we need some more definitions and notation. r_i and p_{ij} are, respectively, the ready time and processing times of job J_i. The due date is d_i, i.e., the promised delivery date of J_i. It is the time by which ideally we would like to have completed J_i.

The allowance for J_i is a_i. It is the period allowed for processing between the ready time and the due date: $a_i = d_i - r_i$.

W_i is the waiting time of J_i preceding its kth operation. By kth operation we do not mean the one performed on M_k (although it may be), but the one that comes kth in the order of processing. If the technological constraints demand that J_i is processed through the machines in the order $M_{j(1)}, M_{j(2)}, M_{j(3)}, \ldots M_{j(m)}$, the kth operation is $o_{ij(k)}$, the one performed on $M_{j(k)}$. W_{ik} is the time that elapses between the completion of J_i on $M_{j(k-1)}$ (or r_i if $k = 1$) and the start of processing on M_k. W_i is the total waiting time of J_i. Clearly,

$$W_i = \sum_{k=1}^{m} W_{ik} \qquad \text{Equ. 6.1}$$

C_i is the completion time of J_i, i.e., the time at which processing of J_i finishes. We have the equality:

$$C_i = r_i + \sum_{k=1}^{m} \left(W_{ik} + p_{ij(k)} \right) \qquad \text{Equ. 6.2}$$

F_i is the flow time of J_i. This is defined to be the time that J_i spends in the workshop. Thus $F_i = C_i - r_i$. L_i is the lateness of J_i. This is the difference between its completion time and its due date:

$L_i = C_i - d_i$. Note that when a job is early, i.e., when it completes before its due date, L_i is negative. It is often more useful to have a variable that, unlike lateness, only takes non-zero values when a job is tardy, i.e., when it completes after its due date. Hence we also define the tardiness and, to be comprehensive, the earliness of a job.

T_i is the tardiness of J_i: $T_i = \max\{L_i, 0\}$.
E_i is the earliness of J_i: $E_i = \max\{-L_i, 0\}$.

These quantities for a typical job are illustrated on the Gantt diagram shown in Figure 6.1. There is a possibility of confusion in some of this terminology. English allows the noun 'time' two distinct meanings. It may be used to refer to an instant or to an interval. Thus ready time and completion time both refer to instants in time, whereas processing time, waiting time, and flow time refer to intervals.

Frequently we use the maximum or the mean of these quantities and it helps to have a compact notation for doing this. Let X_i be any quantity relating to J_i. Then we let $\bar{X} = 1/n\sum_{i=1}^{n} X_i$, the average over all the jobs, and $X_{max} = \max\{X_1, X_2, \ldots, X_n\}$, the maximum over all the jobs. For instance, $\bar{F}$ is the mean flow time and C_{max} is the maximum completion time. Next we define the idle time on machine M_j by

$$I_j = C_{max} - \sum_{i=1}^{n} p_{ij} \qquad \text{Equ. 6.3}$$

In order to see that this definition makes sense, note that C_{max} is the time when all processing ceases and $\sum_{i=1}^{n} p_{ij}$ is the total processing time on machine M_j. Their

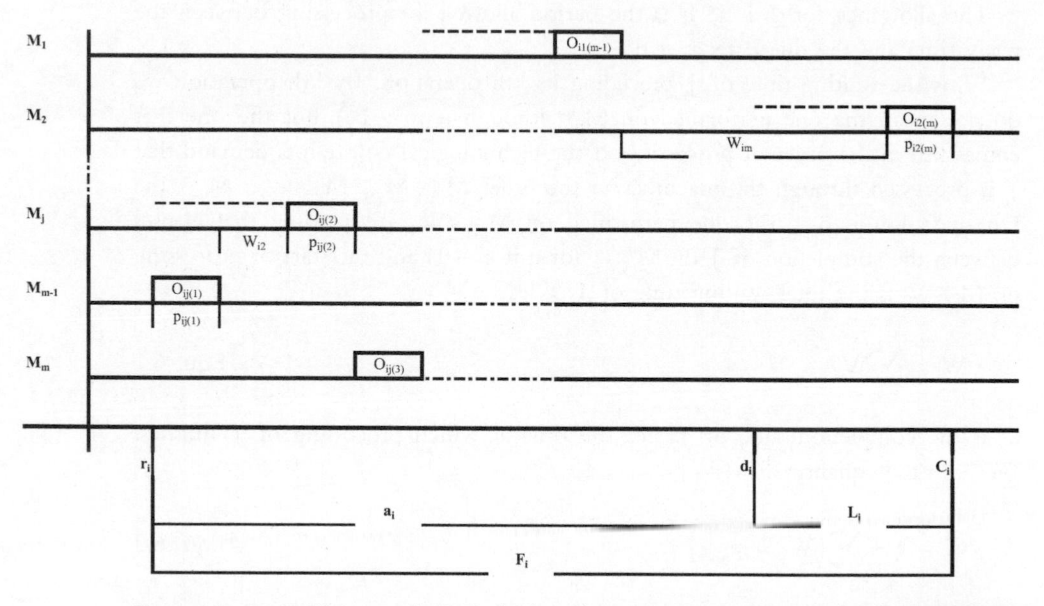

Figure 6.1 Gantt Diagram of a Typical Job J_i. The processing order given by the technological constraints is $(M_{(m-1)}, M_j, M_m, \ldots, M_1, M_2)$. For the waiting times W_{i1}, W_{i3} are zero; W_{i2} and W_{im} are non-zero as shown. For this job $T_i = L_i$ and $E_i = 0$, because the job is completed after its due date.

difference gives the period for which M_j is idle. Now we are in a position to define some measures of performance.

Criteria Based Upon Completion Times

The main criteria in this category are F_{max}, C_{max}, $\bar{F}$, and $\bar{C}$. Minimizing F_{max}, the maximum flow time is essentially saying that a schedule's cost is directly related to its longest job. Minimizing C_{max}, the maximum completion time, says that the cost of a schedule depends on how long the processing system is devoted to the entire set of jobs. Note that in the case where the ready times are zero, C_{max} and F_{max} are identical. When there are non-zero ready times, C_{max} and F_{max} are quite distinct. Indeed, if one job has an extremely late ready time it may easily happen that the job with the shortest flow-time completes at C_{max}. C_{max} is also called the total production time or the makespan. Minimizing $\bar{F}$, the mean flow-time, implies that a schedule's cost is directly related to the average time it takes to process a single job. We shall find that minimizing $\bar{C}$, the mean completion time, is equivalent to minimizing $\bar{F}$; i.e., a schedule that attains the minimum $\bar{C}$ also attains the minimum $\bar{F}$ and vice versa. It may seem strange that F_{max} and C_{max} are quite distinct measures of performance, whereas $\bar{F}$ and $\bar{C}$ are essentially the same. The reason for this is simple. The operation of taking the maximum of a set of numbers has different properties to that of taking the mean.

Criteria Based Upon Due Dates

Because the cost of a schedule is usually related to how much we miss target dates by, obvious measures of performance are $\bar{L}$, L_{max}, $\bar{T}$, and T_{max}; i.e., the mean lateness, the maximum lateness, the mean tardiness, and the maximum tardiness, respectively. Minimizing either L or L_{max} is appropriate when there is a positive reward for completing a job early and that reward is larger the earlier a job is. Minimizing either $\bar{T}$ or T_{max} is appropriate when early jobs bring no reward; there are only the penalties incurred for late jobs.

Sometimes the penalty incurred by a late job does not depend on how late it is; a job that completes a minute late might just as well be a century late. For instance, if an aircraft is scheduled to land at a time after which it will have exhausted its fuel, then the results are just as catastrophic whatever the scheduled landing time. In such cases, a reasonable objective would be to minimize n_T the number or tardy jobs, i.e., the number of jobs that complete after their due dates.

Finally, we note a classification of performance measures into those that are regular and those that are not. A regular measure R is one that is non-decreasing in the completion times. Thus *R* is a function of $C_1, C_2, \ldots, C_n$ such that

$$C_1 \le C_1', \ C_2 \le C_2' \ldots . C_n \le C_n' \text{ together imply that } R(C_1, C_2, .. C_3)$$
$$\le R(C_1', C_2', .. C_n')$$

The rationale underlying this definition is as follows: suppose that we have two schedules such that under the first all the jobs complete no later than they do under the second. Then under a regular performance measure the first schedule is at least as good as the second. Note that we seek to minimize a regular measure of performance. $\bar{C}$, C_{max}, $\bar{F}$, F_{max}, $\bar{L}$, L_{max}, $\bar{T}$, T_{max}, and n_T are all regular measures of performance. For instance, consider the performance measure C_{max} Here we have:

$R(C_1,C_2,\ldots,C_n) = C_{max} = \max(C_1,C_2,\ldots,C_n)$

Let $C_1 \leq C_1'$, $C_2 \leq C_2' \ldots C_n \leq C_n'$. Then

$R(C_1,C_2,\ldots,C_n) = \max(C_1,C_2,\ldots,C_n) \leq \max(C'_1,C'_2,\ldots,C'_n) = R(C'_1,C'_2,\ldots,C'_n)$

Hence C_{max} is a regular performance measure.

Classification of Scheduling Problems

It is convenient to have a simple notation to represent types of job shop problems. We classify problems according to four parameters: n/m/A/B.

n is the number of jobs.

m is the number of machines.

A describes the flow pattern or discipline within the machine shop. When m = 1, A is left blank. A may be:

F for the flow shop case, i.e., the machine order for all jobs is the same.

P for the permutation flow shop case. Here not only is the machine order the same for all jobs, but now we also restrict the search to schedules in which the job order is the same for each machine. Thus a schedule is completely specified by a single permutation of the numbers 1, 2, . . . n, giving the order in which the jobs are processed on each and every machine.

G the general job shop case where there are no restrictions on the form of the technological constraints.

B describes the performance measure by which the schedule is to be evaluated. It may take, for instance, any of the forms discussed in the previous section.

As an example: $n/2/F/C_{max}$ is the n-job, two-machine, flow shop problem where the aim is to minimize the makespan. In using these four parameters we have followed Conway, Maxwell, and Miller (1967), whose definitions have stood the test of time.

Real Scheduling Problems and the Mathematics of the Job Shop

One need not have encountered scheduling problems in practice to realize that they are vastly more complex than those of the job shop as we have defined them. Few obey many, much less all, of the assumptions that we have made. The cost of a schedule is rarely represented well by a function as simple as $\bar{C}$ or T_{max}. You may expect, therefore, that there would be little practical relevance of the theory that we will develop. In fact, our analysis is quite relevant and it is wise to pause and consider why.

First let us examine assumptions 1 to 11 listed earlier. (Assumption 12 differs in nature from the others and is discussed separately.) These assumptions are undoubtedly restrictive. They limit the structure of the job shop problem greatly. They define quite precisely which routings of jobs are allowable and which are not; what the capacities and availabilities of the machines are, and so forth. It is easy to imagine practically occurring problems where some of these restrictions are inappropriate. It is nice to realize that these assumptions are not necessary to the development of a mathematical theory of scheduling; rather they are typical of the assumptions that we can make. The reason for choosing the job shop family is that it leads to a presentation of the theory that is particularly coherent and, furthermore, is not encumbered with a confusion of caveats and provisos needed to cover special cases. Once the job shop family has been studied, it will be an easy matter to follow developments of the theory in other contexts.

Assumption 12 and, to a small extent, Assumption 10, stand distinct from the rest. They confine attention to non-random problems, that is, problems in which all numerical quantities are known and fixed in advance. There is no uncertainty. Because the number of jobs and their ready times are known and fixed, we call our problems *static.* Because the processing times and all other parameters are known and fixed, we call our problems *deterministic.* Problems in which jobs arrive randomly over a period of time are called *dynamic.* Problems in which the processing times, and so forth are uncertain are called *stochastic.* It may be argued that all practical scheduling problems are both dynamic and stochastic, if for no other reason than that all quantities are subject to some uncertainty. In fact, in many problems the randomness is quite obvious. For instance, it may be impossible to predict exactly when jobs will become available for processing, e.g., patients arriving at the emergency room; it may be impossible to predict processing times exactly, e.g., during routine maintenance it will not be known which parts have to be replaced until they have been examined and that examination is one of the operations of the maintenance process; it may be impossible to predict the availability of machines, for some may have significant breakdown rates; and so on. Given that most problems have dynamic and stochastic elements, why do we confine ourselves to static, deterministic cases?

First, there are problems in which any randomness is insignificant because the uncertainty in the various quantities is several orders of magnitude less than the quantities themselves. Because microprocessors and industrial robots have entered production lines, there is a greater degree of certainty in processing times. Second, we cannot study the dynamic and stochastic families of problems until we have understood the static, deterministic family. An awareness of the techniques for scheduling jobs when there is no uncertainty involved will point us toward the solution of stochastic problems. Dynamic stochastic problems are covered in Chapter 15.

So we accept all 12 assumptions, and that leaves us just one more point to discuss here: is our choice of performance measures too limited to allow the representation of the scheduling goals that arise in practice? The first point to note is that in using a performance measure such as $\overline{T}$ we are not assuming that the cost of a schedule is directly proportional to $\overline{T}$, i.e., that the cost is a positive linear function of $\bar{T}$. What we are assuming is that the cost is an increasing function of $\bar{T}$, i.e., a function such that the cost increases when and only when $\bar{T}$ increases; the increases in cost and $\bar{T}$ need not be proportional. Under this condition minimizing $\bar{T}$ minimizes the cost. Thus in restricting the choice of performance measures to those listed earlier, we are not restricting the form of cost function quite as much as we might think.

We are limiting ourselves not because it is wrong to want to minimize, say C_{max}, but because in any real problem we would also want to minimize $\bar{L}$, $\bar{I}$, and so forth. The total cost of a schedule is a complex combination of processing costs, inventory costs, machine idle-time costs, and lateness penalty costs. In other words, each of our performance measures represents only a component of the total cost. A schedule that minimizes a component cost may be (and frequently is) very poor in terms of the total cost.

We lose little by not considering total costs. We will discover that even with simple performance measures scheduling problems can be extremely difficult. Their solution usually requires heuristic or approximate methods (see Chapter 12), and these methods may easily be modified to minimize total costs. Moreover, the insight necessary to enable such modifications will come from our earlier work on problems with simple performance measures.

Optimality of Schedules

The essential problem in scheduling is to time table the processing of jobs by machines so that some measure of performance achieves its optimal value. Two pertinent questions are:

1. For any particular measure of performance does an optimal schedule exist?
2. Given that we have found a schedule that is optimal with respect to one measure, how good is it as compared to a second measure?

Suppose I asked you to find the smallest number strictly greater than 2. If you think about this some of you should see that this apparently simple task is impossible.

If you give me the number 2, then I reply that 2 is not strictly greater than 2. If you give me a slightly greater number, say $(2 + \varepsilon)$ where $\varepsilon > 0$, then I simply point out that $[2 + (\varepsilon/2)]$ is a still smaller number, which is, nonetheless, strictly greater than 2. Thus it is possible to state some minimization problems, for which there is no solution. Clearly for most practical purposes the answer 2.0000000001 to the question is satisfactory, but mathematically it is quite wrong. More importantly, knowledge of whether there is an exact answer to an optimization problem is an important factor in determining our approach to it. If we know there is an optimal solution, then we can search for it directly and, more importantly, test any candidate solution for optimality. Suppose, however, we know that for any solution we can always find one better, albeit only marginally better. Then we must adopt a more practical approach, which seeks a good solution upon which any improvement, although possible, is slight. We show in the next section that in the case of regular measures optimal schedules do exist.

If we have solved a problem for a client, and if she then decides the performance measure was inappropriate and so changes it, do we have to do the work all over again? Earlier we admitted that we would be naïve if we accepted our performance measures as representative of the total costs involved in practical problems; each measure represents only a component of the cost. It would be useful, therefore, to know whether in minimizing one component cost we incidentally minimize another. Schedules that minimize several components simultaneously are generally more satisfactory than those that minimize only one. Therefore we investigate the equivalence of our performance measures.

Regular Measures and Semi-Active Schedules

In Chapter 1, when we were given schedule, we created the associated Gantt chart. We refer to this process as time tabling. Time tabling can also be used to create a Gantt chart when we do not have a schedule. We say that a processing sequence is the order in which the jobs are processed through the machines. A processing sequence, therefore, contains no explicit information about the times at which the various operations start and finish. Figure 1.1 gives an example of a processing sequence. A processing schedule does, however, contain time tabling, as well as sequencing information. Thus the Gantt chart in Figure 1.3 specifies a complete schedule, each block giving the start and finish times of a particular operation. Time tabling is the process whereby we derive a schedule from a sequence.

We note that there are a finite number of distinct processing sequences. Then we show that in the case of regular performance measures we need only consider only one form of time tabling. Thus, once we agree to use this and only this time tabling, each sequence will define a unique schedule. It follows that we only have a finite number of schedules to consider. It is this finiteness that ensures the existence of an optimal schedule. In asking you to find the smallest real number strictly greater than 2, I was asking you to search an infinite set; there are infinitely many numbers slightly greater than 2. Such a search may be unending. But any search of a finite set must end. Here we have a finite set of schedules. So to find one with the smallest value of the performance measure all we need do is search through all the possibilities, comparing one with another until all comparisons have been made and the smallest has been found. Conceptually that is all we need to do. In practice the finite number of schedules may still be too great for such an exhaustive search to be humanly possible. But we only need to show conceptually that an optimal schedule exists.

We begin by noting there are only a finite number of processing sequences. In fact, we have already shown this in Chapter 1 by example. It will do no harm to repeat the argument here. A processing sequence gives for each machine the order of processing jobs. Thus it specifies one permutation of $\{J_1, J_2, \ldots, J_n\}$ for each of the m machines. Because there are n! permutations of $\{J_1, J_2, \ldots, J_n\}$ there are $(n!)^m$ possible processing orders. Actually, because of the technological constraints, many of these may be infeasible. Thus there may be less than $(n!)^m$ processing sequences to consider, but it is clear that the number is finite.

Time tabling is called semi-active if in the resulting schedule there does not exist an operation that could be started earlier without altering the processing sequence or violating the technological constraints and ready dates. In other words, in semi-active time tabling the processing of each operation is started as soon as it can be. If you look back to Figures 1.3, 1.9, and 1.11, you will see that there we used semi-active time tabling without questioning why. It just seemed the sensible thing to do. However, we should question that intuition. Specifically, could Albert and his friends have gone to the park earlier if we had inserted periods of idle time, e.g., if Charles had not picked up the *Los Angeles Times* at 10.00 A.M., but had left it for 5 minutes? Surely not! But we have to justify this.

Theorem 6.1 In order to minimize a regular measure of performance it is only necessary to consider semi-active time tabling.

Proof Consider a schedule that has not been constructed by semi-active time tabling. Then there is at least one operation that could be started earlier. Of all such operations choose one with the earliest finishing time. Time table this operation again to start as early as possible. In this new schedule there cannot be a job whose completion time has increased. Thus the value of a regular measure cannot have increased either.

Repeat this process of time tabling operations again, which could have started earlier, until there are none left. Because we always pick an operation that has the earliest completion time, no operation can be time tabled for a second time more than once. There are a finite number of operations so this time tabling process must terminate. The final schedule is the result of semi-active time tabling, because no operation could be started earlier. Because the value of the performance measure does not increase at any stage, the final schedule is at least as good as the original. Thus for regular measures of performance semi-active time tabling produces schedules at least as good as those we might find by any other method. If you find this proof difficult, you may benefit by drawing yourself a Gantt diagram without using semi-active time tabling and then time tabling operations for a second time as described. In the example of Albert and friends we sought to minimize C_{max}, which is a regular measure.

There are only a finite number of feasible processing sequences. For each processing sequence semi-active time tabling clearly produces a unique schedule. Thus there is only a finite number of these schedules. Theorem 6.1 tells us that in minimizing a regular performance measure we need consider no other schedules. Thus we are minimizing a function over a finite set and so at least an optimal schedule exists. There may, of course, be more than one.

We shall from now on confine ourselves almost entirely to regular measures and, as a consequence of Theorem 6.1, only consider schedules constructed by semi-active time tabling. It follows that a sequence uniquely defines the associated schedule.

Relations Between Performance Measures

We say that two performance measures are equivalent if a schedule that is optimal with respect to one is also optimal with respect to the other and vice versa. Here we prove simple theorems showing that some of the measures introduced earlier in the chapter are equivalent. Let us begin with the simplest.

Theorem 6.2 The following performance measures are equivalent.

(i) $\bar{C}$ (ii) $\bar{F}$ (iii) $\bar{W}$ (iv) $\bar{L}$

Proof From the earlier definitions we have the relations for each job J_i:

$$C_i = F_i + r_i = W_i + \sum_{j=1}^{m} p_{ij} + r_i = L_i + d_i \qquad \text{Equ. 6.5}$$

Summing over the jobs and dividing by n gives the relations between the mean quantities:

$$\bar{C} = \bar{F} + \bar{r} = \bar{W} + (1/n)\sum_{i=1}^{n}\sum_{j=1}^{m} p_{ij} + \bar{r} = \bar{L} + \bar{d} \qquad \text{Equ. 6.6}$$

Now the quantities r_i, $(1/n)\sum_{i=1}^{n}\sum_{j=1}^{m} p_{ij}$, and $\bar{d}$ are constants for each problem and independent of the schedule. Thus in choosing a schedule to minimize $\bar{C}$ we are also minimizing $\bar{F}$, $\bar{W}$, and $\bar{L}$. Similarly minimizing $\bar{F}$ also minimizes $\bar{C}$, $\bar{W}$, and $\bar{L}$, and so on. The four measures are equivalent.

Minimizing the mean completion time of the jobs also minimizes their mean flow time, mean waiting time, and mean lateness. There is no parallel result concerning C_{max}, F_{max}, W_{max}, and L_{max}; these four measures are not equivalent. In the preceding proof the step between equations 6.5 and 6.6 is valid when we take the mean of the quantities involved, but not when we take the maximum. For instance, it is not generally true that $C_{max} = F_{max} + r_{max}$. Consider a two-job, one-machine problem with data:

J_1: $r_1 = 0$, $p_{11} = 5$; J_2: $r_2 = 10$, $p_{21} = 1$

Figure 6.2 shows the schedule that processes J_1 then J_2. Here we have:

$C_1 = 5, C_2 = 11$ so $C_{max} = 11$
$F_1 = 5, F_2 = 1$ so $F_{max} = 5$
$r_1 = 0, r_2 = 10$ so $r_{max} = 10$

and hence we see $C_{max} = 11 \neq 5 + 10 = F_{max} + r_{max}$. Of course, there are special cases when certain of the measures C_{max}, F_{max}, W_{max}, and L_{max} are equivalent. In problems with all ready times zero C_{max} and F_{max} are equivalent (because $C_i = F_i$

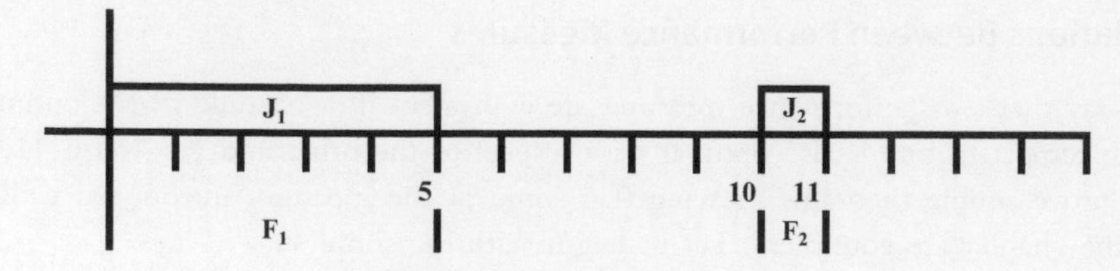

Figure 6.2 Example Showing That C_{max} <> $F_{max} + r_{max}$

for all jobs). Again, in problems where all the due dates are equal to some constant d, C_{max}, and L_{max} are equivalent (because C and L differ by the same constant d for all jobs). Note the partial equivalence of L_{max} and T_{max}.

Theorem 6.3 A schedule that is optimal with respect to L_{max} is also optimal with respect to T_{max}.

Proof

$$T_{max} = \max\{\max\{L_1, 0\}, \max\{L_2, 0\}, \ldots, \max\{L_n, 0\}$$
$$= \max\{L_1, L_2, \ldots, L_n, 0\}$$
$$= \max\{L_{max}, 0\}$$

So minimizing L_{max} also minimizes T_{max}. However minimizing T_{max} does not necessarily minimize L_{max}. Any schedule that finishes all the jobs on or before their due dates has $T_{max} = 0$, its minimum possible value. However, there may be other schedules that finish the jobs even earlier, so reducing L_{max} but leaving $T_{max} = 0$. $\bar{C} / C_{max}$ and $\bar{W} / C_{max}$ are not regular measures of performance, only in the special case of a single-machine problem are they regular—whatever processing sequence we use, C_{max} is a constant, namely the sum of all the processing times.

7 GENERATION OF SCHEDULES

Introduction

In Chapter 1 we saw in the example that it was not easy to come up with a solution, even though the problem had been constructed to allow us to determine the optimal solution. Given about 60 years of research into the scheduling problem, we are reasonably sure that there is no easy way, nor is there likely to be one. Sounds pretty pessimistic, but we nevertheless need to solve problems. The only methods available are those of implicit (or explicit) enumeration, which may take prohibitive amount of computation for problems of practical size (See Chapter 10). If we cannot find the best schedule for the problem within a reasonable time, then we can at least pick one at random because we cannot let a real situation go unscheduled.

We also saw in Chapter 1 that it is not very difficult to come up with a schedule—we can generate one with time tabling or simply make one up and then check to see if it is feasible. What we will not know, however, is how good this schedule is. But one can always say that any schedule is better than no schedule.

We will see in subsequent chapters that the difference in best and worst schedules can be huge, so it is always worthwhile to try and find a better schedule within the time available to us. This time, of course, can range from minutes to days depending on the situation; hence we consider many different methods.

Here we consider algorithms to find better, not necessarily optimal schedules. Most of the time this is a moot point, as no one actually knows what the optimum is and how far any schedule is from the optimum. Note, however, that algorithms here mean prescribed procedures that are not meant to guarantee an optimal solution. These are called heuristic or approximation algorithms. The former term refers to methods justified purely because they seem sensible, that is by 'hand-waving' arguments, and because empirically they are found to perform 'fairly well.' The latter term is occasionally reserved for algorithms that produce solutions guaranteed to be within a certain distance of the optimum (see Chapter 13). The terms, however, are often used interchangeably.

Assumptions for Chapter 7

We make no assumptions to limit the class of problems studied in this chapter, but we do make one strong assumption about the use of heuristic methods. We assume that they are not used when either a constructive, polynomial time solution exists or implicit enumeration is computationally feasible. One should not accept approximations when it is just as easy to find optimal solutions.

Some heuristic methods are applicable to wide classes of problem being essentially approaches to problem solving rather than problem-specific algorithms. Others are ad hoc rules of thumb only applicable to the very specific class of problems for which they were designed.

Schedule Generation Techniques

We begin with a family of methods that spans the spectrum between complete enumeration and heuristic solution. These algorithms are capable of producing all, some, or just one schedule in a particular class. If that class is sure to contain an optimal schedule and if we generate, perhaps implicitly, all that class, then we are again in the realm of optimal solution by enumeration. If, on the other hand, either the class is not guaranteed to contain an optimal schedule or we do not generate all the class, then we may only find an approximate solution. All these possibilities are discussed and clarified in what follows. The methods are applicable to the general job shop: n/m/G/B.

In Chapter 6 we discussed semi-active time tabling. Semi-active schedules ensure that each operation is started as soon as it can be, while obeying both the technological constraints and the processing sequence. Here we consider two further classes of schedules: the active and the non-delay.

In an active schedule the processing sequence is such that no operation can be started any earlier without either delaying some other operation that has already been scheduled or violating the technological constraints. The active schedules form a subclass of the semi-active, i.e., an active schedule is necessarily semi-active. The distinction between active and semi-active schedules may be appreciated by looking back to Chapter 1. Figures 1.3, 1.9, and 1.11 give different schedules for our newspaper-reading friends. All three are semi-active, as was noted there. However, the schedule in Figure 1.3 is not active. Consider the *Financial Times.* Both Charles and Bertrand could be given the paper before Daniel and not cause him any delay; for they would finish it by 11:00 A.M. This has been done in Figure 1.9 and you may check that this is an active schedule. No paper can be given any earlier to anyone without delaying someone else or violating their desired reading order, i.e., the technological constraints. In semi-active time tabling we put the blocks representing operations onto the Gantt chart and slide them to the left as far as we can. In active time tabling we also allow ourselves the possibility of leap-frogging one block over another, provided that there is sufficient idle time waiting to receive it. (Of course, this leap-frogging will change the scheduling sequence that was originally given to the time tabling procedure.)

In a non-delay schedule no machine is kept idle when it could start processing some operation. These schedules form a sub-class of the active. So a non-delay schedule is necessarily active and, hence, necessarily semi-active. Consider the schedule shown in Figure 1.9. This is a non-delay schedule because no paper lies unread when there is someone free to read it.

In Theorem 6.1 we showed that there is necessarily an optimal schedule, which is semi-active. It is easy to modify the proof of this theorem to show that there is necessarily an optimum that is active. Thus we may without loss restrict our attention to the class of active schedules. Because this class is smaller than that of the semi-active, our task of finding an optimal schedule is made easier by this restriction. The class of non-delay schedules is still smaller. However, there need not be an optimal schedule, which is non-delay. This may be seen by considering Figure 1.11, which shows the unique optimum. This schedule is certainly not non-delay. Remember it achieves a final completion time of 11:30 precisely because it insists on Albert's and Bertrand's patience; both the *Financial Times* and the *Guardian* could be being read before 9:05 and 8:50, respectively. Nonetheless, although there is not necessarily an optimal schedule in the class of non-delay ones, we will encounter strong empirical reasons for designing heuristic algorithms that only generate non-delay schedules because non-delay schedules tend to be good.

Next we detail an algorithm that generates some or all of the active schedules for a problem. Then we modify it to generate non-delay schedules. We want to generate an active schedule given the data for an n/m/G/B problem. We could work through a series of semi-active schedules testing each until we found an active one. This would be very time-consuming. A method due to Giffler and Thompson (1960) constructs active schedules right from the beginning.

We require some notation and terminology to describe their algorithm. The algorithm schedules operations one at a time. An operation is schedulable if all those operations that must precede it within its job have already been scheduled. Because there are nm operations, the algorithm iterates through nm stages. At stage t let

P_t—be the partial schedule of the $(t - 1)$ scheduled operations
S_t—be the set of operations schedulable at stage t, i.e., all the operations that must precede those in S_t are in P_t
σ_k—be the earliest time that operation O_k in S_t could be started
φ_k—be the earliest time that operation O_k in S_t could be finished, that is $\varphi_k = \sigma_k + p_k$ where p_k is the processing time of O_k

Algorithm 7.1 (Giffler and Thompson)

Step 1: Let $t = 1$ with P_t being null. S_t will be the set of all operations with no predecessors; namely those that are first in their job.

Step 2: Find $\varphi^\star = \min_{O_k \text{ in } S_k} \{\varphi_k\}$ and the machine $M^\star$ on which $\varphi^\star$ occurs. If there is a choice for $M^\star$, choose arbitrarily.

Step 3: Choose an operation O_k in S_k such that

(1) it requires $M^\star$
(2) $\sigma_j < \varphi^\star$.

There are likely to be several choices, so we can generate a different active schedule for each choice. Because this will be repeated for each iteration, many different (or all) active schedules can be generated.

Step 4: Move to next stage by

(1) Adding O_k to P_t so creating P_{t+1}

(2) Deleting O_k from S_t and creating S_{t+1} by adding to S_t the operation that directly follows O_k in its job (unless O_k completes its job)

(3) Incrementing t by 1.

Step 5: If there are any operations left unscheduled ($t < nm$), go to step 2. Otherwise, stop.

Let us check that this algorithm produces an active schedule, that is, one in which no operation could be started earlier without delaying some other operation or breaking the technological constraints. At each stage the next operation to schedule is chosen from S_t so the algorithm can never break the technological constraints. Hence to prove that an active schedule is generated we must show that to start any operation earlier must delay the start of some other. With this in mind let us work through the algorithm.

Step 1 initializes t, P_t, and S_t in the obvious fashion. So suppose that we come to step 2 at stage t with $0 \le t < mn$ and with P_t and S_t set appropriately by the earlier stages. $\varphi^\star$ gives the earliest possible completion time of the next operation added to the partial schedule P_t.

Next consider the machine $M^\star$ on which $\varphi^\star$ is attained. (We may choose $M^\star$ arbitrarily if there is a choice) In step 2, we examine all schedulable operations that need $M^\star$ and can start before $\varphi^\star$.

There is at least one such operation, namely one that completes at $\varphi^\star$. Of these we select one and schedule it to start as soon as it can. The method of selection here will be the subject of much discussion shortly. We have selected an operation O_j for $M^\star$ that can start strictly before $\varphi^\star$. Moreover, if we try to start any of the other operations schedulable on $M^\star$ before the selected operation O_j we know that they must complete at some time $\ge \varphi^\star$. Hence they must delay the start of O_j. We see, therefore, that this algorithm is steadily building up an active schedule. Step 4 moves from one stage to the next, while step 5 determines whether the complete schedule has been generated. Note that for any active schedule there is a sequence of choices at step 3 that will lead to its generation.

As an example of the use of the algorithm we generate one active schedule for the newspaper reading problem of Chapter 1. In doing so we use the notation (A, F) to represent Albert's reading of the *Financial Times,* and similarly for the other operations. For convenience the data are repeated in Figure 7.1. Figure 7.2 gives

Reader	Gets up at	Reading Order and Times in Minutes			
Albert	8:30	F	L	E	N
		60	30	2	5
Bertrand	8:45	L	E	F	N
		75	3	25	10
Charles	8:45	E	L	F	N
		5	15	10	30
Daniel	9:30	N	F	L	E
		90	1	1	1

Figure 7.1 Data for the Newspaper Example

	Paper next free at:								Also	operation scheduled	
t	F	L	E	N	Operation	σ_k	φ_k	φ^*	Eligible	o_j	Reason
1	8:30	8:30	8:30	8:30	(A, F)	8:30	9:30				
					(B, L)	8:45	10:00				
					(C, E)	8:45	8:50	8:50		(C, E)	Only one
					(D, N)	9:30	11:00				
2	8:30	8:30	8:50	8:30	(A, F)	8:30	9:30				
					(B, L)	8:45	10:00	10:00		(B, L)	MWKR
					(C, L)	8:50	9:05		(C, L)		
					(D, N)	9:30	11:00				
3	8:30	10:00	8:50	8:30	(A, F)	8:30	9:30	9:30		(A, F)	Only one
					(B, E)	10:00	10:03				
					(C, L)	10:00	10:15				
					(D, N)	9:30	11:00				
4	9:30	10:00	8:50	8:30	(A, L)	10:00	10:30				
					(B, E)	10:00	10:03	10:03		(B, E)	Only one
					(C, L)	10:00	10:15				
					(D, N)	9:30	11:00				
5	9:30	10:00	10:03	8:30	(A, L)	10:00	10:30		(A, L)		
					(B, F)	10:03	10:28				
					(C, L)	10:00	10:15	10:15		(C, L)	SPT
					(D, N)	9:30	11:00				
6	9:30	10:15	10:03	8:30	(A, L)	10:15	10:45				
					(B, F)	10:03	10:28		(B, F)		
					(C, F)	10:15	10:25	10:25		(C, F)	SPT
					(D, N)	9:30	11:00				
7	10:25	10:15	10:03	8:30	(A, L)	10:15	10:45	10:45		(A, L)	Only one
					(B, F)	10:25	10:50				
					(C, N)	10:25	10:55				
					(D, N)	9:30	11:00				
8	10:25	10:45	10:03	8:30	(A, E)	10:45	10:47	10:47		(A, E)	Only one
					(B, F)	10:25	10:50				
					(C, N)	10:25	10:55				
					(D, N)	9:30	11:00				
9	10:25	10:45	10:47	8:30	(A, N)	10:47	10:52				
					(B, F)	10:25	10:50	10:50		(B, F)	Only one
					(C, N)	10:25	10:55				
					(D, N)	9:30	11:00				
10	10:50	10:45	10:47	8:30	(A, N)	10:47	10:52	10:52	(A, N)		
					(B, N)	10:50	11:00		(B, N)		
					(C, N)	10:25	10:55		(C, N)		
					(D, N)	9:30	11:00			(D, N)	MWKR
11	10:50	10:45	10:47	11:00	(A, N)	11:00	11:05				
					(B, N)	11:00	11:10				
					(C, N)	11:00	11:30				
					(D, F)	11:00	11:01	11:01		(D, F)	Only one
12	11:01	10:45	10:47	11:00	(A, N)	11:00	11:05				
					(B, N)	11:00	11:10				
					(C, N)	11:00	11:30				
					(D, L)	11:01	11:02	11:02		(D, L)	Only One
13	11:01	11:02	10:47	11:00	(A, N)	11:00	11:05				
					(B, N)	11:00	11:10				
					(C, N)	11:00	11:30				
					(D, E)	11:02	11:03	11:03		(D, E)	Only one
14	11:01	11:03	10:47	11:00	(A, N)	11:00	11:05	11:05		(A, N)	SPT
					(B, N)	11:00	11:10		(B, N)		
					(C, N)	11:00	11:30		(C, N)		
15	11:01	11:03	11:03	11:05							
					(B, N)	11:05	11:15	11:15		(B, N)	SPT
					(C, N)	11:05	11:35		(C, N)		
16	11:01	11:03	11:03	11:15							
					(C, N)	11:15	11:45	11:45		(C, N)	Only one

Figure 7.2 Iterations to Generate One Active Schedule for the Newspaper Example

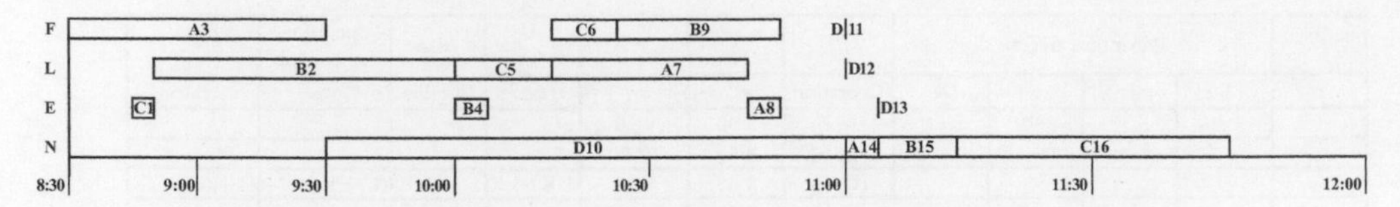

Figure 7.3 Gantt Chart for the Active Schedule for the Newspaper Example

the calculations needed to generate each iteration of the active schedule. At each stage any choice at step 3 of the algorithm has been made according to a randomly chosen rule (see the decision rules later in the chapter). The calculations in the figure are straightforward. It should be noted that the scheduling of an operation in step 3 not only has implications for the earliest moment when the paper concerned is free at the next stage, but also will change the earliest start and finish times of other operations requiring that paper. Thus, when Charles is given the *Los Angeles Times* at stage 5, the earliest time that Albert can begin to read it drops from 10:00 to 10:15. The resulting schedule is shown in the Gantt chart of Figure 7.3. The numbers after the person indicate the sequence in which the algorithm placed the jobs into the chart.

When modifying the layout of Figure 7.2 for other scheduling problems it should be remembered that there the papers are the machines and the readers are the jobs.

It is straightforward to modify this algorithm so that it produces non-delay schedules. Steps 2 and 3 are the only ones that are different. Instead of considering the operation with the earliest finishing time, we consider the one with the earliest start time.

Algorithm 7.2

Step 1: Let $t = 1$ with P_t being null. S_t will be the set of all operations with no predecessors; in other words, those that are first in their job.

Step 2: Find $\sigma^\star = \min_{O_k \text{ in } S_k}\{\sigma_k\}$ and the machine M★ on which σ★ occurs. If there is a choice for M★, choose arbitrarily.

Step 3: Choose an operation O_k in S_k such that

(1) it requires M★
(2) $\sigma_j = \sigma^\star$

Step 4: Move to next stage by

(1) adding O_k to P_t so creating P_{t+1}
(2) deleting O_k from S_t and creating S_{t+1} by adding to S_t the operation that directly follows O_k in its job (unless O_k completes its job)
(3) incrementing t by 1.

Step 5: If there are any operations left unscheduled ($t < nm$), go to step 2. Otherwise, stop.

Neither algorithm is fully defined; there is an undefined choice at step 3. To this we now turn.

We could pass through either algorithm many times creating all possible sequences resulting from all choices at step 3. We would then have generated all the active or all the non-delay schedules. Because the set of active schedules is guaranteed to contain an optimum, generating all and selecting the best leads to an optimal solution. There are often far fewer active schedules than semi-active; so this process of complete enumeration may be feasible for small problems. Such an approach rapidly becomes computationally infeasible as the problem size grows. Some respite may be gained by sacrificing optimality and only generating non-delay schedules. There are fewer of these generally and empirically it is known that the best non-delay schedule, if not optimal, is certainly very good. But even this is infeasible for problems of moderate size.

An obvious progression from complete enumeration is to consider implicit enumeration. Thus these algorithms may be used as a basis for a branch and bound solution. (See Chapter 10.) For most problems we cannot avoid the problem of making the choice at step 3 simply by making all possible choices. Because this leads to very large problems, we need to find a way to reduce the number of choices. Many selection or priority rules have been discussed in the literature. The most studied of these are the six listed here.

SPT (shortest processing time) Select the operation with the shortest processing time.

FCFS (first come first served) Select the operation that has been in S_t for the greatest number of stages.

MWKR (most work remaining) Select the operation that belongs to the job with the greatest total processing time remaining.

LWKR (least work remaining) Select the operation that belongs to the job with the least total processing time remaining.

MOPNR (most operations remaining) Select the operation that belongs to the job with the greatest number of operations still to be processed.

RANDOM (random) Select an operation at random.

The reasons for suggesting these rules are not hard to see. Using the SPT rule finds an operation that monopolizes the machine M* for the least time and, hence, perhaps constrains the whole system least. The FCFS rule is based upon a notion of fairness and is perhaps more applicable in queuing theory, from which it is derived. The MWKR rule is based upon the notion that it may be best not to get too far behind on a particular job; whereas the LWKR rule stems from the contrary notion that perhaps one should concentrate on finishing some jobs first in the hope that the later jobs may be speeded through a less cluttered system. Finally, the feeling that delays in scheduling derive from the changeover of jobs between machines suggests that we should hurry through the jobs with the most changeovers left, hence the MOPNR rule.

Of course, even the application of one of these rules may not fully define the operation to schedule next. For instance, there may be two operations satisfying the conditions of step 3 and also sharing the same shortest processing time of all such operations. The SPT rule cannot, therefore, choose between these. Thus it is not sufficient to decide to use just one of these rules to select the scheduled operation at step 3. Rather one must decide upon a hierarchy of rules. Perhaps first use the MWKR rule; if this does not choose a unique operation, choose among those operations selected by the MWKR by the SPT rule; if this does not choose a unique operation, resolve the remaining ambiguity by choosing arbitrarily. However, note first that the RANDOM rule selects a unique operation; and second that the RANDOM rule is, of course, equivalent to choosing arbitrarily.

As an example, let us apply each of these priority rules to stage 10 of the active schedule generation given in Figure 7.2. There are four operations that might be scheduled at step 3, namely (A, N), (B, N), (C, N) and (D, N); each requires the *New York Times* and each may start before $\varphi^{\star}= 10{:}52$. Here:

the SPT rule selects (A, N), because Albert reads the *New York Times* the fastest;
the FCFS rule selects (D, N), because (D, N) has been schedulable since stage 0;
the MWKR rule selects (D, N), because Daniel has the most reading left;
the LWKR rule selects (A, N), because Albert only has 5 minutes of reading left;
the MOPNR rule selects (D, N), because Daniel still has to read all four papers;
the RANDOM rule selects one of (A, N), (B, N), (C, N) and (D, N) at random.

Jeremiah, Lalchandani, and Schrage (1964) have examined the performance of these and other priority rules empirically. Their results are summarized well in Conway, Maxwell, and Miller (1967, pp. 121–124) and hence are mentioned only briefly here. The authors generated both active and non-delay schedules under each of these priority rules for some 84 different problems. In well over 80% of the cases the non-delay schedules performed better in terms of mean flow time than the active schedules. For minimizing maximum flow time the non-delay schedules were again superior, but not as markedly. The SPT, RANDOM, and LWKR rules clearly dominated the rest for minimizing mean flow time, although it must be noted that there were problems in which some other priority rule was superior. For minimizing maximum flow time the MWKR rule was a clear winner with SPT a distant second. See also Gere (1966).

Because we have mentioned that non-delay schedules tend to be very good, we should do an example of it as well. When I generated the preceding active schedule I had no idea what the outcome would be, although I used rules that made sense. We will do the same with the non-delay schedule. See Figures 7.4 and 7.5. We should note here that the completions are in the figure only as a guide to drawing the Gantt start—they are not instrumental in generating the non-delay schedule.

Notice that both methods generated the same maximum completion time and were worse than the optimum. You should also notice that in generating the active schedule we made six non-arbitrary choices, while in generating the non-delay schedule we made three non-arbitrary choices. This is a good indication that there are fewer non-delay schedules, although it is hard to tell how many subsequent choices there will be in the processes. The reason that we only count non-arbitrary choices is that arbitrary choices do not lead to different results.

	Paper next free at:							Also	operation scheduled		
t	F	L	E	N	Operation	σ_k	σ^*	Eligible	o_j	Reason	φ_k
1	8:30	8:30	8:30	8:30	(A, F)	8:30	8:30		(A, F)	Only one	9:30
					(B, L)	8:45					10:00
					(C, E)	8:45					8:50
					(D, N)	9:30					11:00
2	9:30	8:30	8:30	8:30	(A, L)	9:30					10:00
					(B, L)	8:45	8:45		(B, L)	Arbitrary	10:00
					(C, E)	8:45	8:45				8:50
					(D, N)	9:30					11:00
3	9:30	10:00	8:30	8:30	(A, L)	10:00					10:30
					(B, E)	10:00					10:03
					(C, E)	8:45	8:45		(C, E)	Only one	8:50
					(D, N)	9:30					11:00
4	9:30	10:00	8:50	8:30	(A, L)	10:00					10:30
					(B, E)	10:00					10:03
					(C, L)	10:00					10:15
					(D, N)	9:30	9:30		(D, N)	Only one	11:00
5	9:30	10:00	8:50	11:00	(A, L)	10:00	10:00				10:30
					(B, E)	10:00	10:00		(B, E)	arbitray	10:03
					(C, L)	10:00	10:00				10:15
					(D, F)	11:00					11:01
6	9:30	10:00	10:03	11:00	(A, L)	10:00	10:00				10:30
					(B, F)	10:03					10:28
					(C, L)	10:00	10:00		(C, L)	SPT	10:15
					(D, F)	11:00					11:01
7	9:30	10:15	10:03	11:00	(A, L)	10:15					10:45
					(B, F)	10:03	10:03		(B, F)	Only one	10:28
					(C, F)	10:15					10:25
					(D, F)	11:00					11:01
8	10:28	10:15	10:03	11:00	(A, L)	10:15	10:15		(A, L)	Only one	10:45
					(B, N)	11:00					11:10
					(C, F)	10:28					10:38
					(D, F)	11:00					11:01
9	10:28	10:45	10:03	11:00	(A, E)	10:45					10:47
					(B, N)	11:00					11:10
					(C, F)	10:28	10:28		(C, F)	Only one	10:38
					(D, F)	11:00					11:01
10	10:38	10:45	10:03	11:00	(A, E)	10:45	10:45		(A, E)	Only one	10:47
					(B, N)	11:00					11:10
					(C, N)	11:00					11:30
					(D, F)	11:00					11:01
11	10:38	10:45	10:47	11:00	(A, N)	11:00	11:00				11:05
					(B, N)	11:00					11:10
					(C, N)	11:00					11:30
					(D, F)	11:00			(D, F)	Arbitray	11:01
12	10:38	10:45	10:47	11:00	(A, N)	11:00	11:00		(A, N)	SPT	11:05
					(B, N)	11:00		(B, N)			11:10
					(C, N)	11:00		(C, N)			11:30
					(D, L)	11:01					11:02
13	10:38	10:45	10:47	11:05							
					(B, N)	11:05					11:15
					(C, N)	11:05					11:35
					(D, L)	11:01	11:01		(D, L)	Only one	11:02
14	10:38	10:45	10:47	11:05							
					(B, N)	11:05					11:15
					(C, N)	11:05					11:35
					(D, E)	11:02	11:02		(D, E)	Only one	11:03
15	10:38	10:45	11:03	11:05							
					(B, N)	11:05	11:05		(B, N)	SPT	11:15
					(C, N)	11:05		(C, N)			11:35
16	10:38	10:45	11:03	11:15							
					(C, N)	11:15	11:15	(C, N)		Only one	11:45

Figure 7.4 Iterations to Generate One Non-Delay Schedule for the Newspaper Example

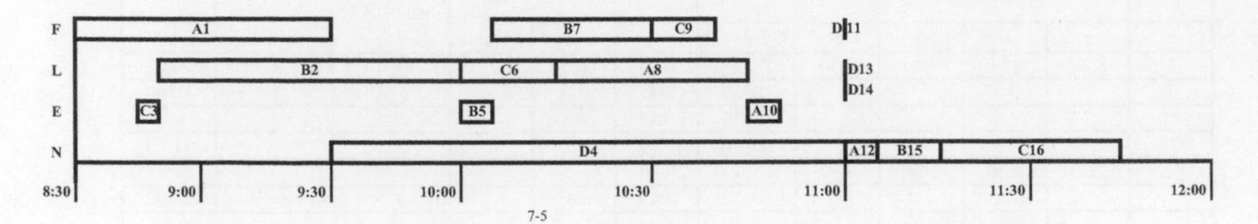

Figure 7.5 Gantt Chart for the Non-Delay Schedule for the Newspaper Example

Finally we turn to probabilistic dispatching or Monte Carlo methods. So far in this discussion we have considered either enumeration of all the schedules within a class or, at the other extreme, the generation of just one schedule through the choice of a hierarchy of priority rules. In between these extremes we may generate a sample of schedules within the class and then pick the best of these. We might, for instance, use the RANDOM priority rule, but, instead of generating just one schedule, we might pass through the algorithm, say, 50 times thus producing a sample of schedules. However, in some sense it seems wrong to choose the operation at step 3 simply at random, that is with equal probability from the possible choices. Some choices are both intuitively and empirically better than others. Were this not so, we would not have suggested and discussed the five non-random priority rules listed earlier. Yet, if we replace the RANDOM rule with one of these and generate 50 schedules, we shall find that we generate the same schedule 50 times—a rather pointless exercise. What we can do is bias the probabilities used in the random selection to favor those operations that seem the most sensible choice. Thus, returning to stage 10 of our active example, we might randomly choose between the operations (A, N), (B, N), (C, N) and (D, N) according to, say, the probabilities 4/12, 0/12, 0/12, and 8/12, respectively, because we know that empirical evidence favors the MWKR rule for minimizing maximum flow-time. Biasing the probabilities like this at each stage means that in generating a sample of schedules we will tend to find those, in some sense, near the unique schedule produced by the deterministic application of the MWKR rule.

We shall follow up this method of probabilistic dispatching no further here except to say that with a sensible choice of biasing probabilities it does seem to work well, substantially better than the deterministic application of a priority rule. See Chapter 12 for a further exploration of Monte Carlo methods. Both Conway, Maxwell, and Miller (1967) and Baker (1974) discuss these procedures in much greater depth and report on empirical results.

8 ALGORITHMS FOR ONE-MACHINE PROBLEMS

Introduction

We are now in a position to begin solving problems, not particularly difficult ones, as this chapter is only concerned with the simplest ones involving just one machine. We stop classifying, discussing existence and equivalence of solutions, and attempt to solve something.

We shall assume throughout this chapter that all jobs are ready for processing at the beginning of the processing period. This is not a very restrictive condition—we are usually only interested in scheduling jobs that are ready. Different situations are dealt with in later chapters.

Assumptions for Chapter 8

1. $r_i = 0$ for all J_i $\quad i = 1, 2, \ldots, n$
2. $m = 1$

Generally in each chapter we make a number of assumptions over and above those listed in Chapter 6. These will apply for the duration of that chapter alone and we list them explicitly in the introduction.

It is interesting to note that single-machine scheduling problems arise in practice more often that one might expect. First, there are obvious ones involving a single machine, e.g., the processing of jobs through a small, non-time-sharing computer. Then, there are less obvious ones where a large complex plant acts as if it were one machine, e.g., in paint manufacture the whole plant may have to be devoted to making one color of paint at a time. Finally, there are job shops with more than one machine, but in which one machine acts as a 'bottleneck,' e.g., in the hospital example of Chapter 6 the treatment of patients may be severely restricted by the shortage of theater facilities. Thus M_3 the actual surgical operation may determine the rate of treatment completely, the other hospital facilities easily coping with the demands put on them. It makes sense to tackle such m-machine problems with single bottlenecks as single-machine problems, if only to get a first approximation to their solution. Because there is only one operation per job, we shall denote by p_i the processing time of job J_i, i.e., we drop the '1' from our earlier notation of p_{i1}.

Permutation Schedules

In the example of Albert and his friends we saw that certain schedules, in particular the one in Figure 1.3, cause the machines, i.e., the papers, to be idle when perhaps they needed not to be. For instance, Albert could have started to read the *Financial Times* at 8:30, but he leaves it idle so that Charles may have it first. This tactic, not processing the jobs immediately available, but waiting for one that will shortly be so, may be employed by the optimal schedule in the general n/m/A/B problem. However, we shall now see that it is unnecessary to consider such inserted idle time for a single-machine problem with a regular measure of performance. In fact, our result follows immediately from Theorem 6.1. There is no harm in proving it again from first principles and doing so will emphasize a style of argument that is common to much of our subject.

Theorem 8.1 For an n/1//B problem, where B is a regular measure of performance, there exists an optimal schedule in which there is no inserted idle time, i.e., the machine starts processing at $t = 0$ and continues without rest until $t = C_{max}$.

Proof By the argument in Chapter 6 we know an optimal schedule exists (B is a regular measure). Let S be an optimal schedule with inserted idle time from $t = \tau_1$ to $t = \tau_2$. Let S′ be the schedule obtained from S by bringing forward by an amount $(\tau_2 - \tau_1)$ all operations that start after τ_1. Note that S′ is feasible because there is only one machine. Clearly the completion times C_i under S′ are such that $C'_i \leq C_i$ for $i = 1, 2, . . ., n$, where the C_i are the completion times under S. Thus the value of any regular measure cannot increase by moving from S to S′. Then S′ must also be optimal. This process may be repeated to remove all inserted periods of idle time, leaving an optimal schedule in which processing is continuous.

In Chapter 6 we made the assumption that operations could not be preempted; i.e., we are not allowed to interrupt the processing of one operation by the processing of a second before returning to complete that of the first. For the general n/m/A/B problem this is a constraint on our solution; we might be better with a schedule involving preemption than we can do with the best one that does not. However, in the single-machine case with a regular measure of performance this is not so; there is no advantage gained by using preemption.

Theorem 8.2 In an n/1//B problem, where B is a regular measure of performance, no improvement may be gained in the optimal schedule by allowing preemption.

Proof Suppose we have an optimal schedule S in which job I is preempted to allow job K to start and then completes without preemption at some later time. See the Gantt diagram in Figure 8.1. We assume K completes without preemption. Let S′ be the schedule obtained from S by interchanging the first part of I with K. Then

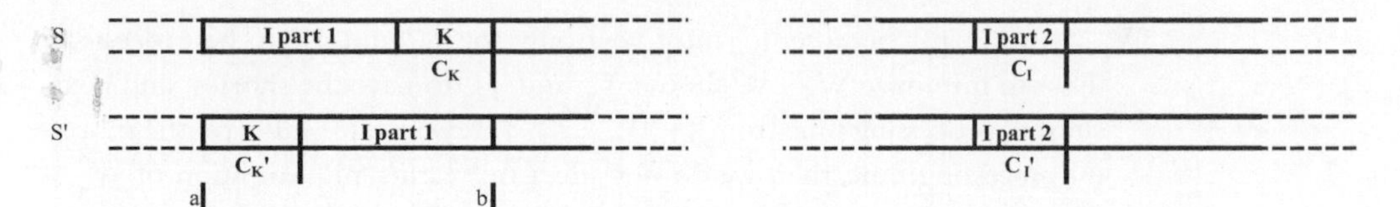

Figure 8.1 Gantt Diagram of Schedules S and S′

no completion times outside the interval [a, b] differ between S and S′. Indeed there is only one change at all in completion time: $C'_k < C_k$

Hence the value of a regular measure cannot increase by moving from S to S′. S is optimal, so S′ is also. By repeating this process the first part of job I can be leap-frogged down the schedule until it is continuous with its last part. Hence there exists an optimal schedule in which job I is not preempted. It follows that there exists an optimal schedule entirely without preemption.

The significance of Theorems 8.1 and 8.2 is that in n/1//B problems with B regular we need only consider permutation schedules. In other words, our task is to find a permutation of the jobs $J_1, J_2, \ldots, J_n$ such that, when they are sequenced in that order the value of B is minimized. Note there are n! permutations of n jobs, so finding the best one could be a formidable task, if not for the algorithms we are about to cover.

We shall write $J_{i(k)}$ for the job is scheduled at the kth position in the processing sequence and use the subscript i(k) appropriately. Thus J_i is a generic job drawn from the list of jobs $\{J_1, J_2, \ldots, J_n\}$ and $J_{i(k)}$ is the job that the processing sequence selects as the kth to be processed, k = 1,2, . . ., n.

Shortest Processing Time Scheduling

Suppose we wish to minimize mean flow time, i.e., our problem is an $n/1//\overline{F}$. For a particular processing sequence:

$$\overline{F} = \frac{1}{n}\sum_{i=1}^{n}(W_i + p_i) = \frac{1}{n}\sum_{k=1}^{n}\left(W_{i(k)} + p_{i(k)}\right) = \frac{1}{n}\sum_{k=1}^{n}W_{i(k)} + \sum_{k=1}^{n}p_{i(k)}$$

(on ordering the sum as in the processing sequence)

Now $\sum_{k=1}^{n} p_{i(k)} = \sum_{i=1}^{n} p_i$ is a constant for all sequences. Hence to minimize $\overline{F}$ we must minimize $\sum_{k=1}^{n} W_{i(k)}$ If we choose a sequence to make each $W_{i(k)}$ as small as it could possibly be, then we shall clearly minimize their sum.

$W_{i(1)} = 0$ for all sequences, because $J_{i(1)}$ starts immediately.
$W_{i(2)} = p_{i(1)}$ because $J_{i(2)}$ must wait only for $J_{i(1)}$ to be processed. Thus if we choose $J_{i(1)}$ to have the shortest processing time of all the jobs $\{J_1, J_2, \ldots, J_n\}$, we shall minimize $W_{i(2)}$.

$W_{i(3)} = p_{i(1)} + p_{i(2)}$ because $J_{i(3)}$ must wait only for $J_{i(1)}$ and $J_{i(2)}$ to be processed. Thus to minimize $W_{i(3)}$ we choose $J_{i(1)}$ and $J_{i(2)}$ to have the shortest and next shortest processing time from $\{J_1, J_2, \ldots, J_n\}$. If we let $J_{i(1)}$ still have the shortest processing time, then we do not affect our earlier minimization of $W_{i(2)}$. Therefore we can minimize $W_{i(3)}$ and $W_{i(2)}$ simultaneously.

Continuing in this way we build up a schedule in which at the kth job to be processed has the shortest processing time of those remaining. Doing so simultaneously minimizes all the waiting times. Thus the result is an SPT schedule (shortest processing time) that minimizes $\overline{F}$. We state this as Theorem 8.3, to which we also provide an alternative proof. The reason for proving this theorem again is that the alternative proof has a form common to many of our later arguments.

Theorem 8.3 For an $n/1//\overline{F}$ problem, the mean flow time is minimized by sequencing such that $p_{i(1)} \leq p_{i(2)} \leq p_{i(3)} \leq \ldots \leq p_{i(n)}$ where $p_{i(k)}$ is the processing time of the job that is processed kth.

Proof Let S be a non-SPT schedule. Then for some k

$$p_{i(k)} > p_{i(k+1)} \qquad \text{Equ. 8.1}$$

Let S′ be the schedule obtained by interchanging $J_{i(k)}$ and $J_{i(k+1)}$ in S (see Figure 8.2).

For convenience, label $J_{i(k)}$ as I and $J_{i(k+1)}$ as K. All jobs other than I and K have the same flow times in S′ as in S. So the difference in mean flow time for S and S′ depends only on the flow times for jobs I and K.

$$\text{Let } a = \sum_{i=1}^{k-1} p_{i(k)} = \begin{Bmatrix} W_I \textit{ in } S \\ W_k \textit{ in } S' \end{Bmatrix}$$

So in S, $F_I = a + p_I$ and $F_K = a + p_I + p_K$. Similarly in S′, $F'_I = a + p_K + p_I$ and $F'_K = a + p_K$.

Hence the contribution of I and K to $\overline{F}$ in S is greater than their contribution to $\overline{F'}$ in S′.

$$\frac{1}{n}(F_I + F_K) = \frac{1}{n}(2a + p_I + p_k + p_I) > \frac{1}{n}(2a + p_I + p_k + p_k) = \frac{1}{n}\left(F_I' + F_K'\right)$$

because $p_I > p_k$

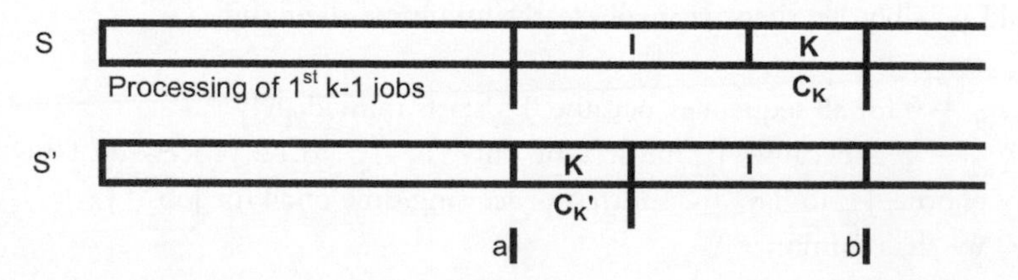

Figure 8.2 Gantt Diagram of Schedules S and S′

Thus S′ has smaller mean flow time than S and so any non SPT schedule can be bettered. Therefore an SPT schedule must solve an $n/1//\overline{F}$ problem. If two or more jobs have equal processing times there will be more than one such schedule. Unless it is otherwise stated, such ties in our algorithms always allow for an arbitrary choice because either choice will result in the same optimal measure. From the equivalences shown earlier we also note that SPT scheduling solves the following problems: $n/1//\overline{C}, n/1//\overline{W}, n/1//\overline{L}$, i.e., it also minimize the mean completion time, the mean waiting time, the mean lateness, but not mean tardiness.

Example Given the following $7/1//\overline{F}$ problem (Figure 8.3):

Job	1	2	3	4	5	6	7
Processing Time	2	5	8	3	2	9	1

Figure 8.3 SPT Example Problem

To find an optimal SPT schedule we reorder the jobs in non-decreasing order of processing times and calculate the total flow time by adding the processing time of each succeeding job to the flow time of the previous one (Figure 8.4):

Job	7	5	1	4	2	3	6	
Processing Time	1	2	2	3	5	8	9	Total
Completion time	1	3	5	8	13	21	30	81

Figure 8.4 Solution to the SPT Example Problem

Because we have two jobs with equal processing times, there are also two optimal schedules (7154236 and 7514236). Also note that for comparison purposes there is no need to divide by n, as it is a constant. It is also instructive to see how bad the schedule could be by looking at the worst one (Figure 8.5).

Job	6	3	2	4	5	1	7	
Processing Time	9	8	5	3	2	2	1	Total
Completion time	9	17	22	25	27	29	30	159

Figure 8.5 Worst Solution to the SPT Example Problem

There are n! = 7! = 5040 possible sequences. The distribution of sums of flow times for these is shown in Figure 8.6. You can see that the algorithm is very efficient in finding the best ones.

The expression for the average flow time is:

$$\overline{F} = \frac{1}{n}\sum_{k=1}^{n}(n-k+1)p_{i(k)} \qquad \text{Equ. (8.2)}$$

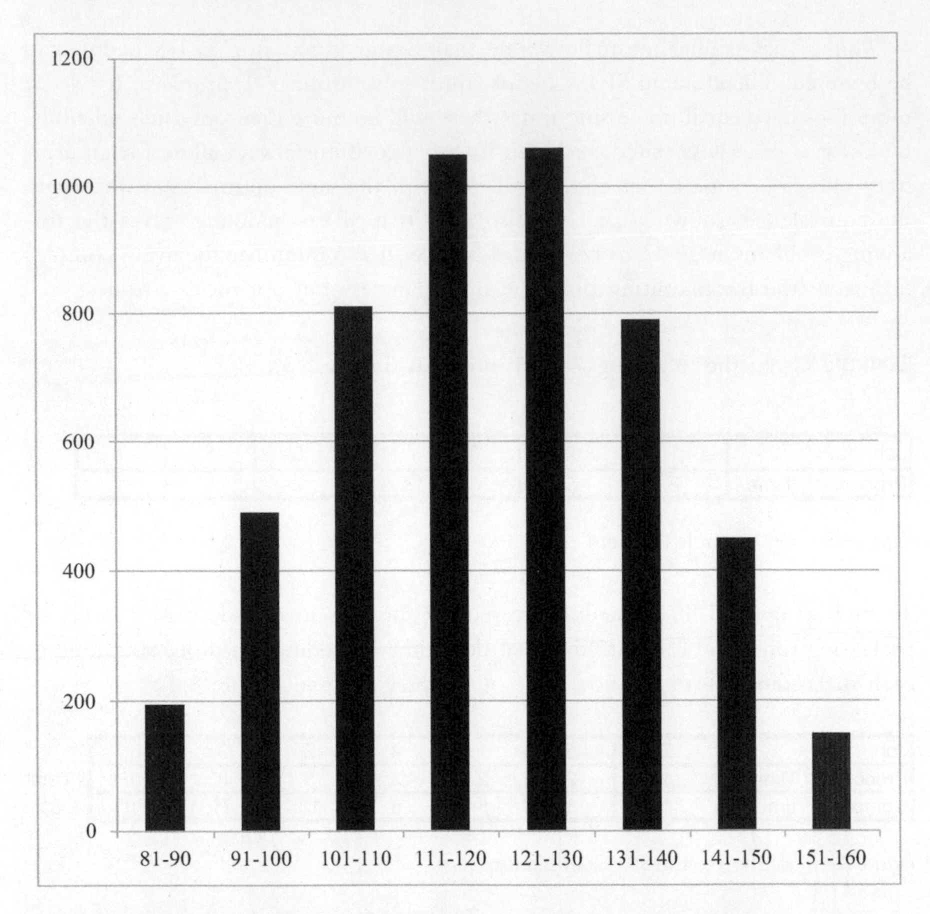

Figure 8.6 Distribution of Sum of Flow Times for the SPT Example Problem

Earliest Due Date Scheduling

A reasonable approach to scheduling would be to sequence jobs in the order in which they are required. In other words, to sequence the jobs such that the first processed has earliest due date, the second processed the next earliest due date, and so on. What does this accomplish? The answer is given by Theorem 8–4.

Theorem 8.4 For an $n/1//L_{max}$ problem, the maximum lateness is minimized by sequencing such that $d_{i(1)} \leq d_{i(2)} \leq d_{i(3)} \leq \ldots \leq d_{i(n)}$ where $d_{i(k)}$ is the due date of the job that is processed kth.

Proof Suppose S is a schedule in which the jobs are not ordered according to increasing due date. Then, as for the previous theorem, for some k, $d_{i(k)} > d_{i(k+1)}$, label jobs $J_{i(k)}$ and $J_{i(k+1)}$ as I and K respectively. So we have $d_I > d_K$. Let S′ be the schedule obtained by interchanging jobs I and K and leaving the rest of the sequence unaltered. You should look back to Figure 8.2 in the proof of Theorem 8.3. Let L be the maximum lateness of the (n – 2) jobs other than I and K under S and L′ be the corresponding quantity under S′. Clearly L = L′. Let L_I, L_K be the lateness of I and K under S and L'_I, L'_K be the corresponding quantities under S′. We, therefore have the maximum lateness under S L_{max} = max(L, L_I, L_K), and the maximum lateness under S′ L'_{max} = max(L′, L'_I, ‘L_K),= max(L, L'_I, ‘L_K), because L = L′.

Under S: $L_I = a + p_I - d_I$; $L_K = a + p_I + p_K - d_K$, where $a = \sum_{l=1}^{k-1} p_{I(l)}$
and under S′: $L'_I = a + p_I + p_K — d_I$; $L'_K = a + p_K - d_K$
Therefore $L_K > L'_K$ (because $p_I > 0$) and $L_K > L'_I$ because $d_I > d_K$
Hence $L_K > \max(L'_I, L'_K) \Rightarrow \max(L, L_I, L_K) \geq \max(L, L_K)$
$\max(L, L'_I, L'_K) = \max(L', L'_I, L'_K)$ and finally $L_{max} \geq L'_{max}$

Thus any schedule can be rearranged into increasing due date (actually non-decreasing because there can be equal due dates) order without increasing its maximum lateness and the theorem is proven.

By Theorem 6.3 we note that we have also solved the $n/1//T_{max}$ problem. This method of sequencing is called earliest due date (EDD).

Example Given the $7/1//T_{max}$ problem in Figure 8.7:

Job	1	2	3	4	5	6	7
Processing Time	1	2	2	3	4	5	3
Due Date	12	20	8	6	17	9	21

Job	4	3	6	1	5	2	7
Processing Time	3	2	5	1	4	2	3
Due Date	6	8	9	12	17	20	21
Completion time	3	5	10	11	15	17	20
Tardy	0	0	1	0	0	0	0

Figure 8.7 Example Problem for Maximum Lateness and Its Solution

We generate an optimal EDD sequence by arranging the jobs in non-decreasing order. In order to calculate the maximum lateness or tardiness we generate the completion times as we did in the SPT example and get max(0, $C_i - d_i$) for each job and take the maximum of these (T_{max} = 1). At this point you should also use SPT to reorder the jobs and find that SPT gives T_{max} = 11. It is important to remember that obtaining an optimum for one measures generally does not get us an optimum for another measure unless the measures are equivalent. The worst schedule has T_{max} = 14. The distribution for all 5040 schedules follows in Figure 8.8.

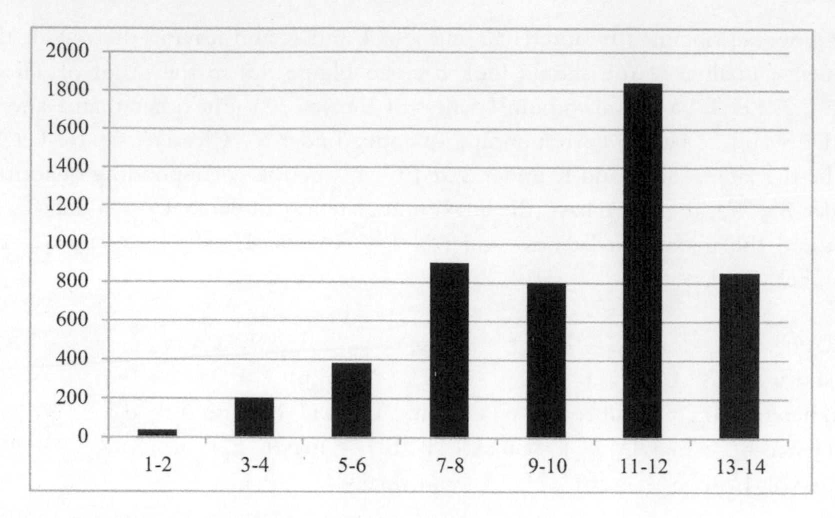

Figure 8.8 Distribution of T_{max} for the EDD Example

Moore's Algorithm

It occasionally makes sense to penalize tardy jobs equally, however late they are. Thus it would cost just as much to miss a due date by 1 week as by a hundred years. For instance, Fisher and Jaikumar (1978) use this kind of penalty in scheduling space shuttle flights, because to miss a launch date even by a few hours can completely upset a space mission. If we adopt this philosophy, our scheduling task is to minimize the number of tardy jobs; i.e., we face an $n/1//n_T$ problem. We consider an algorithm for solving this due to Moore, but in a form suggested by Hodgson. We first state the algorithm and give an example of its use. Then we prove that it does find an optimal schedule.

Algorithm 8.1 (Moore and Hodgson)

Step 1: Sequence the jobs according to the EDD rule to find the current sequence $(J_{i(1)}, J_{i(2)}, \ldots, J_{i(n)})$ such that $di_{(k)} \leq d_{i(k+1)}$ for $k = 1, 2, \ldots, n-1$

Step 2: Find the first tardy job, say $J_{i(p)}$ in the current sequence. If no such job is found, go to step 4.

Step 3: Find the job in the sequence $(J_{i(1)}, J_{i(2)}, \ldots, J_{i(p)})$ with the largest processing time and reject this from the current sequence. The effect of doing this is to lower the completion time of all subsequent jobs as much as possible, thus increasing their chances of not being tardy. Return to step 2 with a current sequence that is one shorter than the one before.

Step 4: Form an optimal schedule by taking the current sequence and appending to it the rejected jobs, which may be sequenced in any order. If there is a secondary objective not involving n_T, these rejected jobs may of course be resequenced to satisfy a secondary criterion. The rejected jobs will all be tardy and these will be the only tardy jobs.

Example Given the $8/1//n_T$ problem and the subsequent steps to solve it in Figure 8.9. We first form the EDD sequence and compute the completion times until a tardy job is found (steps 1 and 2). Job 1 is the first tardy job in the sequence and of the subsequence (4, 3, 7, 1) job 7 has the largest processing time. So we reject job 7 (step 3), not job 1! We return and repeat step 2 with the new current sequence. We find no more tardy jobs and conclude the $n_T = 1$ and the sequence (4, 3, 1, 6, 2, 8, 5) is optimal. An exploration of all 8! = 40,320 schedules shows that there are actually 150 optimal sequences and that the worst sequence has $n_T = 7$ and of which there are 12. Once again we can see that an algorithm has found a much better schedule than what you would get at random. In Figure 8.10 you can also see that the poor schedules vastly outnumber the good ones.

Job	1	2	3	4	5	6	7	8
Processing Time	2	2	2	3	4	4	5	3
Due Date	11	20	8	6	23	17	10	21

Job	4	3	7	1	6	2	8	5
Processing Time	3	2	5	2	4	2	3	4
Due Date	6	8	10	11	17	20	21	23
Completion time	3	5	10	12	16	14	17	21
Tardy	0	0	0	1	0	0	0	0

Job	4	3	1	6	2	8	5
Processing Time	3	2	2	4	2	3	4
Due Date	6	8	11	17	20	21	23
Completion time	3	5	7	11	9	12	16
Tardy	0	0	0	0	0	0	0

Figure 8.9 n_T Example

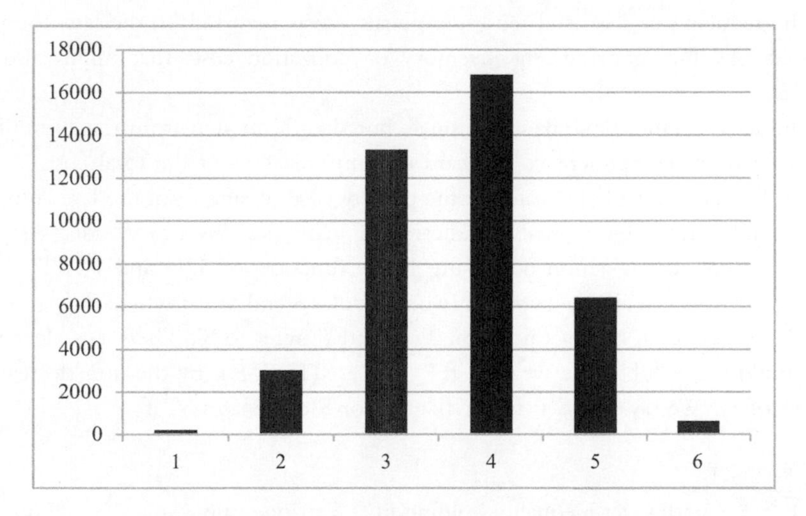

Figure 8.10 Distribution of n_T in the Moore Example

Precedence Constraints and Efficiency

Often one does not consider all of one's customers to be of equal importance. Consequently we will want to process one job before another. In this chapter we assume that all our jobs are to be processed on one machine. The fact that one job has to be processed before another is referred to as a precedence constraint.

Another example of a precedence constraint is when one job is a prerequisite or component part of another job. The existence of precedence constraints in a problem is both a complication and a simplification. It may actually help in the solution of a problem because it reduces the number of feasible sequences. But this does not mean that you should think of precedence constraints as a trick by which we may avoid particularly difficult problems. In very many cases they enter a problem quite naturally. For instance, in a computing example it is not because of cost that we run a second program after the first; it is simply because of a logical necessity. The output of one is the input of the other.

It is understandable that some people should confuse technological constraints and precedence constraints. Technological constraints give the order in which the operations that compose each job must be processed. In other words, they give the routing that each job must follow through the machines. You cannot tap a hole before you drill it. For a single machine there is only one operation per job, hence there can be no technological constraints. Precedence constraints restrict the sequence of processing operations between different jobs. These require that a certain operation of one job must be fully processed before a certain operation of a different job may be begun.

In this chapter we consider two simple classes of single-machine problems with precedence constraints. In a later chapter we further discuss the form that precedence constraints may take.

Later in this chapter we turn from precedence constraints to efficiency. For instance, if we seek to minimize the maximum tardiness of a schedule, then we will be reducing in a general sense the penalty costs incurred by the late completion of jobs, but ignoring any inventory or utilization costs that might also be incurred.

The idea of efficiency takes us a small, but significant step from the minimization of a single component cost toward the minimization of the total cost.

Let T_{max} be a suitable indicator of the penalty costs arising from the late completion of jobs and $\overline{F}$ be a suitable indicator of in-process inventory costs. Assume that the total cost is a non-decreasing linear function of T_{max} and $\overline{F}$, $f(T_{max}, \overline{F})$. Suppose we have to choose between two schedules S and S′ for which $T_{max} < T_{max}'$, $\overline{F} < \overline{F}'$, where in an obvious notation T_{max}' and $\overline{F}'$ refer to S′. Then it is clear that we should prefer S because we have $f(T_{max}, \overline{F}) < f(T_{max}', \overline{F}')$ (by the non-decreasing nature of f). We say that S is better than S′, or S dominates S′, if

$$T_{max} \leq T_{max}'$$
$$\overline{F} \leq \overline{F}', \text{ with strict inequality holding in at least one case.} \qquad \text{Equ. 8.3}$$

We say that a schedule S′ is efficient if there does not exist a schedule S that dominates it, i.e., such that Equation 8.3 holds. To be strictly correct, when saying a schedule is efficient, we should also state the performance measures concerned. In general, a schedule S′ is efficient with respect to measures $\mu_1, \mu_2, \ldots, \mu_n$ if there is no such schedule S as: $\mu_1 \leq \mu_1'$; $\mu_2 \leq \mu_2'$ ……. $\mu_n \leq \mu_n'$ with strict inequality holding in at least one case, when $\mu_1, \mu_2, \ldots, \mu_n$ are not equivalent measures.

If we can identify the set of efficient schedules with respect to performance measures $\mu_1, \mu_2, \ldots, \mu_n$ then we may have made a useful step toward finding the schedule that minimizes total cost. For if the total cost is an increasing function of $\mu_1, \mu_2, \ldots, \mu_n$ alone, the minimum cost schedule must be efficient. Later we return to our example above and for a single-machine problem find the set of schedules that are efficient with respect to T_{max} and $\bar{F}$. Having found this efficient set, we determine an optimal schedule for a particular linear total cost function.

Lawler's Algorithm

This algorithm was developed by Lawler (1973) to deal with general precedence constraints. Let's look at the situation where we have to process certain jobs before, but not necessarily immediately before, certain others. Lawler's algorithm minimizes the maximum cost of processing a job, where this cost has a general form $\gamma_i(C_i)$ for J_i and is assumed to be non-decreasing in the completion time C. The usefulness of the general form lies in the fact that frequently different jobs are evaluated with different objectives, even if they are processed on the same machine. For example, one job is judged by how much it contributes to inventory while another is judged by its lateness. The algorithm minimizes

$$\max_{i=1\ to\ n}\{\gamma_i(C_i)\} \qquad \text{Equ. 8.4}$$

Because of the non-decreasing nature of each $\gamma_i(C_i)$, it follows immediately that this performance measure is regular. At first sight Equation 8.4 looks somewhat foreboding compared with our earlier performance measures; however, for specific choices of $\gamma_i(C_i)$, this equation takes on much more familiar forms. If $\gamma_i(C_i) = C_i - d_i = L_i$ then Equation 8.4 gives the measure L_{max}; if $\gamma_i(C_i) = \max\{C_i - d_i, 0\}$, it gives T_{max}. To develop the algorithm we need the following theorem.

Theorem 8.5 Consider the $n/1//\max_{i=1\ to\ n}\{\gamma_i(C_i)\}$ problem with precedence constraints. Let V denote the subset of jobs that may be performed last, i.e., those jobs that are not required to precede any other. Note that the final job in the schedule must complete at $\tau = \sum_{i=1}^{n} p_i$. Let J_k be a job in V such that

$$\gamma_k(\tau) = \min_{J_i\ in\ V}\{\gamma_i(\tau)\} \qquad \text{Equ. 8.5}$$

That is, of all the jobs that may be performed last, J_k incurs the least cost. Then there is an optimal schedule in which J_k is scheduled last.

We now deduce the form of the algorithm. Let J_k be the job that by Theorem 8.5 may be last in an optimal sequence. Thus there is an optimal sequence of the form (A, J_k), where A is a permutation of the other (n – 1) jobs. The maximum cost of this sequence (A, J_k) is the larger of $\gamma_k(\tau)$, the cost of completing J_k last, and the maximum cost of completing the jobs in A. An optimal sequence can be found by making both these terms as small as possible. J_k has been chosen so that $\gamma_k(\tau)$ is the minimum for all the jobs that could be processed last. So to construct an optimal sequence our task is to choose A so that the maximum cost of completing its jobs is as small as possible. In other words, we face the problem of scheduling (n – 1) jobs subject to precedence constraints so that the maximum cost of the individual jobs is minimized. We face a new problem with the same form as our original, but with (n – 1) jobs instead of n. By Theorem 8.5 we can say which job should be last for this new problem and, hence, (n – 1)st in the sequence for the original problem. We are left with the task now of scheduling (n – 2) jobs. And so we go on, repeatedly scheduling a job at the end of the sequence and reducing the size of the problem by 1. Eventually we completely solve the original problem.

Algorithm 8.2 (Lawler)

Step 1: Set k = n
Set $\tau = \sum_{i=1}^{n} p_i$

Step 2: Select the set V of jobs eligible to finish last (having no followers)

For each of these jobs determine the cost of finishing last. Select job i from these with the smallest cost and place it in position k of the eventual sequence.
Set k = k – 1
If k =1, place the remaining job in V in position 1 and stop
Set $\tau = \tau$ – processing time of job i
Reduce the set V by job i
Go to step 2

Example Consider the $6/1//L_{max}$ problem with data and the constraints that J_1 must precede J_2, which must in turn precede J_3, and also that J_4 must precede both J_5 and J_6. These are shown in Figure 8.11.

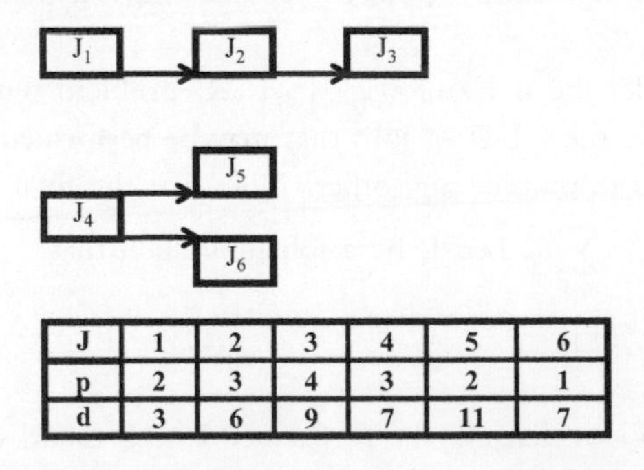

J	1	2	3	4	5	6
p	2	3	4	3	2	1
d	3	6	9	7	11	7

Figure 8.11 Precedence Constraints and Data for the L_{max} with Precedences Example

Finding the job processed sixth: $\tau = 2 + 3 + 4 + 3 + 2 + 1 = 15$. Jobs J_3, J_5 and J_6 can be processed last, i.e., $V = \{J_3, J_5, J_6\}$ the minimum lateness over $V = \min\{(15 - 9), (15 - 11), (15 - 7)\}$, which occurs for J_5. therefore J_5 is scheduled 6th.

Finding the job processed fifth: We delete J_5 from our list and note that the completion time of the first five jobs is $\tau = 15 - 2 = 13$. Jobs J_3 or J_6 can be processed last now; i.e., $V = \{J_3, J_6\}$. So the minimum lateness over $V = \min\{(13 - 9), (13 - 7)\}$, which occurs for J_3. So job J_3 is scheduled 5th.

Finding the job processed fourth: We have now deleted J_3 and J_5 from our list. Thus J_2 becomes available for processing last. We have $\tau = 13 - 4 = 9$ and $V = \{J_2, J_6\}$ minimum lateness over $V = \min\{(9 - 6), (9 - 7)\}$, which occurs for J_6. Thus J_6 is scheduled 4th.

Finding the job processed third: J_3, J_5 and J_6 have now been deleted; hence we now have $V = \{J_2, J_4\}$ and $\tau = 9 - 1 = 8$. Minimum lateness over $V = \min\{(8 - 6), (8 - 7)\}$, which occurs for J_4, thus J_4 is scheduled third.

The jobs scheduled first and second are now automatically J_1 and J_2, respectively, because of the precedence constraints. These calculations are laid out concisely in Figure 8.12.

The entries in column J_i are respectively: * if J_i cannot be scheduled last, or $\tau - d_i$ if it is possible to schedule J_i last or S if J_i has already been scheduled.

The scheduled job at each τ has the minimum lateness $\tau - d_i$ and this value is shaded in the table. The final schedule may be found by reading up the final column and the value of L_{max} found by taking the largest of the shaded quantities. Here we find the schedule (J_1, J_2, J_4, J_6, J_3, J_5) with $L_{max} = 4$. You should note that the algorithm generates a sufficient optimal sequence but not a necessary one, as there are 6 sequences of the 40 feasible ones that are also optimal (see Figure 8.13).

τ	J_1	J_2	J_3	J4	J5	J6	Scheduled Job
15	*	*	6	*	4	8	J_5
13	*	*	4	*	S	6	J_3
9	*	3	S	*	S	2	J_6
8	*	2	S	1	S	S	J_4
5	*	-1	S	S	S	S	J_2
2	-1	S	S	S	S	S	J_1

Figure 8.12 Solution Table for the L_{max} with Precedences Example

6	1	2	3	5
1	6	2	3	5
1	2	6	3	5
4	6	2	3	5
4	2	6	3	5
2	4	6	3	5

Figure 8.13 Optimal and Feasible Sequences for the L_{max} Problem with Precedences Example

Schedules Efficient With $\overline{F}$ and T_{max}

We now consider how to find schedules that are efficient with respect to $\overline{F}$ and T_{max}. Here we describe an algorithm due to Smith (1956) that, although not directed to the immediate solution of the problem, is nonetheless instrumental in the construction of efficient schedules. We follow Smith in the motivation of his theory.

Suppose that we have a $n/1//T_{max}$ problem to solve. We know that a solution may be found by the EDD rule. Suppose that when we construct this schedule we find that $T_{max} = 0$, i.e., that all the due dates can be met. Then, before we actually process the jobs in EDD sequence, perhaps we should stop and think. There may be schedules other than the EDD one that also satisfy the due date constraints. Would we prefer one of these? For instance, one may give a smaller value of $\overline{F}$ than the EDD sequence. Smith's algorithm gives us a way of finding a schedule to minimize $\overline{F}$ subject to the condition that $T_{max} = 0$.

As does Lawler's, Smith's algorithm builds up a schedule from back to front; it first finds a job to be nth in the processing sequence, then $(n - 1)$st, and so on. Again as with Lawler's, it is based upon a theorem that describes the characteristics of a job that may be processed last.

Theorem 8.6 For the n-job, one-machine problem such that all the due dates can be met, there exists a schedule that minimizes $\overline{F}$ subject to $T_{max} = 0$ and in which job J_k is last, if and only if

(i) $d_k \geq \sum_{i=1}^{n} p_i$
(ii) $p_k \geq p_i$ for all jobs J_i such that $d_i \geq \sum_{i=1}^{n} p_i$

Proof Consider a schedule S in which job J_k is last and conditions (i) and (ii) hold. Suppose we interchange J_k and some job J_l in the sequence. If $d_l < \sum_{i=1}^{n} p_i$, i.e., if condition (i) does not hold for job J_l then J_l will be tardy and so T_{max} will no longer be zero. On the other hand, if $d_l \geq \sum_{i=1}^{n} p_i$, but $p_k > p_l$, i.e., if condition (ii) does not hold for the new sequence, then we shall show that the mean flow time of the sequence is increased. Let the schedule with J_k and J_l interchanged by S′. Then, using a prime to indicate quantities connected with S′, we have:

$$F_k' > F_l \text{ because } p_k > p_l$$
$$F_l' = F_k = \sum_{i=1}^{n} p_i$$

And the flow time of any job scheduled between J_k and J_l will be increased by $p_k - p_l > 0$. The flow times of all earlier jobs will be unchanged. It follows immediately that $\overline{F}' > \overline{F}$. The theorem is proven.

Algorithm 8.3 (Smith)

For this algorithm we use the following notation:

k is the position in the processing sequence currently being filled (k cycles down to n, (n −1), . . .,1);

τ. is the time at which the job kth in the sequence must complete; and U is the set of unscheduled jobs

Step 1: Set k = n, $\tau = \sum_{i=1}^{n} p_i$; U={$J_1$, J_2, . . .,J_n}

Step 2: Find $J_{i(k)}$ in U such that (i) $d_{i(k)} > \tau$ and (ii) $p_{i(k)} \geq p_l$, for all J_i in U such that $d_l \geq \tau$

Step 3: Decrease k by 1; decrease τ by $p_{i(k)}$; delete $J_{i(k)}$ from U

Step 4: If there are more jobs to schedule, i.e., if k ≥ 1, go to step 2.

Otherwise stop with the optimal processing sequence ($J_{i(1)}$, $J_{i(2)}$ ..$J_{i(n)}$).

There is no need to first check whether T_{max} = 0. If there is no schedule with T_{max} = 0, we shall discover this in working through Smith's algorithm. At some pass through step 2 we shall be unable to find any $J_{i(k)}$ with $d_{i(k)} > \tau$. Second, sometimes $J_{i(k)}$ may not be determined uniquely; there may be two or more jobs for which conditions (i) and (ii) are true. If this is so, we make the choice of $J_{i(k)}$ arbitrarily—at least for the present.

Example Solve the 4/1/ / $\bar{F}$ problem subject to T_{max} = 0 for the data in Figure 8.14.

J	1	2	3	4
p	2	3	1	2
d	5	6	7	8

Figure 8.14 Data for the Smith Example

Applying the algorithm:

Step 1: k = 4. τ = 8, U = {J_1 J_2, J_3, J_4}.

Step 2: Only J_4 satisfies condition (i) so we choose $J_{i(4)} = J_4$.

Step 3: k = 3; τ = 6; U = {J_1 J_2, J_3}.

Step 4: k ≥ 1.

Step 2: J_2 and J_3 satisfy condition (i); J_2 has the larger processing time, so $J_{i(3)} = J_2$.

Step 3: k = 2; τ = 3; U = {J_1, J_3}.

Step 4: k ≥ 1.

Step 2: J_1 and J_3 satisfy condition (i); J_1 has the larger processing time, so $J_{i(2)} = J_1$.

Step 3: k = 1; τ = 1; U = {J_3}.

Step 4: k ≥ 1.

Step 2: J_3 satisfies condition (i) so $J_{i(3)} = J_3$.

Step 3: k = 0; τ = 1; U is empty.

Step 4: One optimal sequence is (J_3, J_1, J_2, J_4).

Feasible Sequences				Total Flow
2	1	3	4	22
3	1	2	4	18
1	3	2	4	19
1	2	3	4	21

Figure 8.15 Feasible Sequences for the Smith Example

Note that the sequence has $\overline{F} = 18/4$, whereas the EDD sequence (J_1. J_2, J_3, J_4) has $\overline{F} = 21/4$. Thus we have the example promised earlier that shows that the EDD sequence need not minimize $\overline{F}$ subject to $T_{max} = 0$. In fact, there are four sequences (Figure 8.15) that satisfy $T_{max} = 0$.

Smith left his theory here, but Van Wassenhove and Gelders (1980) have developed his ideas further. The rationale underlying Smith's approach is that we are only willing to consider reducing $\overline{F}$ once we have ensured that $T_{max} = 0$. In other words, penalty costs are the most important. Yet, if we are prepared to allow T_{max} to rise, we might be able to reduce $\overline{F}$ more than sufficiently to compensate. So let us turn our attention to minimizing $\overline{F}$ subject to $T_{max} \leq \Delta$, i.e., subject to no job being finished more than Δ after its due date. (Some customers may be willing to tolerate some tardiness in exchange for lower costs resulting from lower inventory).

The key to our analysis is realizing that minimizing $\overline{F}$ subject to $T_{max} \leq \Delta$ in one problem is equivalent to minimizing $\overline{F}$ subject to $T_{max} \leq 0$ in another related problem in which all the due dates have been increased by Δ. A schedule has $T_{max} \leq \Delta$ in the first problem if and only if the same schedule has $T_{max} = 0$ in the second. Thus we may solve our $n/1//\overline{F}$ problem subject to $T_{max} \leq \Delta$ by adding Δ to all the due dates and applying Algorithm 8.3 to the modified problem.

Finding Schedules Efficient With Respect to T_{max} and $\overline{F}$

We now turn to the question of finding efficient schedules. A schedule is efficient with respect to T_{max} and $\overline{F}$ if we can find no other schedule at least as good in both criteria and strictly better in at least one (see the conditions in Equation 8.3). We consider the schedules constructed by Smith's algorithm. Are these efficient? If we minimize $\overline{F}$ subject to, $T_{max} \leq \Delta$ do we find an efficient schedule? The answer is that we may.

Consider what happens at step 2 of the algorithm. More than one job may satisfy conditions (i) and (ii). For now we make an arbitrary choice of the job to process kth. Now a schedule that minimizes $\overline{F}$ subject to $T_{max} \leq \Delta$ may have $T_{max} \leq \Delta$. Moreover, the arbitrary choice at step 2 may determine the final value of T_{max}. This suggests that, when two or more jobs satisfy condition (i) and (ii), we need to make a rather more careful choice. To argue what form this choice should take we need two observations:

(a) If we interchange two jobs not necessarily adjacent in the processing sequence, but both with the same processing time, then $\overline{F}$ is unchanged.

(b) Again suppose that there are two not necessarily adjacent jobs with the same processing time, but now suppose also that the job earlier in the sequence has the later due date. Then interchanging these two jobs cannot increase T_{max} and may decrease it.

We now consider the choice that may face us at step 2 of Smith's algorithm. Look at any two jobs I and K satisfying (i) and (ii). By condition (ii) they must have the equal processing times. Suppose we choose job I and suppose that K has a later due date. Now K will be scheduled by some later cycle of the algorithm and so will be placed earlier in the processing sequence than I. By observation (a) we may interchange I and K in the processing sequence without affecting $\bar{F}$. Moreover, by observation (b) this interchange may decrease T_{max}. Thus it would have been better to select job K when making the choice of step 2. Hence we modify Smith's algorithm to ensure this choice. Step 2 is replaced by:

Step 2: Find $J_{i(k)}$ in U such that (i) $d_{i(k)} > \tau$ and (ii) $p_{i(k)} \geq p_l$, for all J_l in U such that $d_l \geq \tau$. If there is a choice for $J_{i(k)}$ choose $J_{i(k)}$ to have the largest possible due date.

When this modification is made, the algorithm always finds an efficient schedule.

Theorem 8.7 Suppose that we use Smith's algorithm with the modified step 2 to solve an $n/1//\bar{F}$ subject to $T_{max} \leq \Delta$ problem. Then either the algorithm fails to construct a schedule, i.e., it is impossible to have $T_{max} \leq \Delta$, or the resulting schedule is efficient with respect to T_{max} and $\bar{F}$. (The algorithm is applied to the problem after all the due dates have been increased by Δ.)

Proof We stated earlier that the algorithm will fail to construct a schedule if the constraint upon T_{max} is impossible. Suppose then it constructs a schedule S′ with values T_{max}' and $\bar{F}'$. Let S be any other schedule with values T_{max} and $\bar{F}$. We consider three cases $\bar{F} < \bar{F}'$, $\bar{F} = \bar{F}'$, $\bar{F} > \bar{F}'$.

Suppose $\bar{F} < \bar{F}'$. Because $\bar{F}'$ is the minimum mean flow time subject to the restriction on the maximum tardiness, we must have $T'_{max} \leq \Delta < T_{max}$. Hence S does not dominate S′. Suppose $\bar{F} = \bar{F}'$. First note that Theorem 8.6 gives an if and only if condition. So Smith's original algorithm can, in principle, find all the schedules that minimize mean flow time subject to the restriction on the maximum tardiness. Our modification to step 2 ensures that of such schedules S′ has the least possible maximum tardiness. Thus we must have $T'_{max} \leq T_{max}$ and again S does not dominate S′. Suppose $\bar{F} > \bar{F}'$ Then clearly S does not dominate S′. In no case does S dominate S′. Hence S′ must be efficient.

Note that modification to Smith's algorithm may still leave us with a choice to make at step 2. There may be two or more jobs all satisfying conditions (i) and (ii) and all sharing the same due date. In such cases we can choose arbitrarily without any risk of upsetting the efficiency of the resulting schedule.

Our next task is to find all the efficient schedules for a problem. To do this we make one further assumption, namely that all the processing times and due dates are integral. Now for all possible schedules $T_{max} \leq \sum_{i=1}^{n} p_i$, because no job can be more tardy than the total processing time. Thus we may find all efficient schedules by minimizing $\overline{F}$ subject to $T_{max} \leq \Delta$ for $\Delta = 0, 1, 2, \ldots, \sum_{i=1}^{n} p_i$. We need only consider integral values of Δ because we have assumed that the problem's data are integral. There are two qualifications that we should make. First, we shall only find all efficient schedules if, whenever step 2 involves an arbitrary choice between jobs with equal due dates, we repeat the solution until all possible sets of arbitrary choices have been made. Each repetition will lead to a distinct efficient schedule. Second, there is no need to solve the problems for all values of Δ. Suppose that in minimizing $\overline{F}$, subject to $T_{max} \leq \Delta_1$, we obtain a schedule with $T_{max} = \Delta_o < \Delta_1$ then this same schedule will also minimize $\overline{F}$ subject to $T_{max} \leq \Delta$ for all Δ such that $\Delta_o \leq \Delta \leq \Delta_1$ thus we are led to the following algorithm that generates all the efficient schedules for our problem.

Algorithm 8.4 (Van Wassenhove and Gelders)

Step 1: Set $\Delta = \sum_{i=1}^{n} p_i$.

Step 2: Solve the n/1// $\overline{F}$ subject to $T_{max} \leq \Delta$ using the modified version of Smith's algorithm. If step 2 of that algorithm involves an arbitrary choice, repeat the solution until all possible choices have been made. If there is no schedule with $T_{max} \leq \Delta$, go to step 5.

Step 3: Let the schedule(s) found in step 2 have $T_{max} \leq \Delta_o$. Set $\Delta = \Delta_o - 1$.

Step 4: If $\Delta \geq 0$, go to step 2. Otherwise, continue to step 5.

Step 5: Stop.

Example Find all efficient schedules for the four-job problem with the data in Figure 8.16 and an objective function $f = 4T_{max} + 3\overline{F}$:

Note that $\sum F$ and $\overline{F}$ are interchangeable in an optimization problem, as the division by a constant does not change the choice of solution.

Applying Algorithm 8.4 we obtain:

Step 1: $\Delta = 10$.

Step 2: We find the efficient sequence (J_4, J_1, J_3, J_2) with $\sum F = 20$ and $T_{max} = 8$
$f = 4(8) + 3(20)/4 = 47$.

Step 3: $\Delta = 8 - 1 = 7$.

Step 4: $\Delta \geq 0$.

Step 2: We find the efficient sequence (J_4, J_1, J_2, J_3) with $\sum F = 21$ and $T_{max} = 6$
$f = 4(6) + 3(21)/4 = 39.75$.

Step 3: $\Delta = 6 - 1 = 5$.

Step 4: $\Delta \geq 0$

Step 2: We find the efficient sequence (J_1, J_2, J_3, J_4) with $\sum F = 27$ and $T_{max} = 5$
$f = 4(5) + 3(27)/4 = 40.25$.

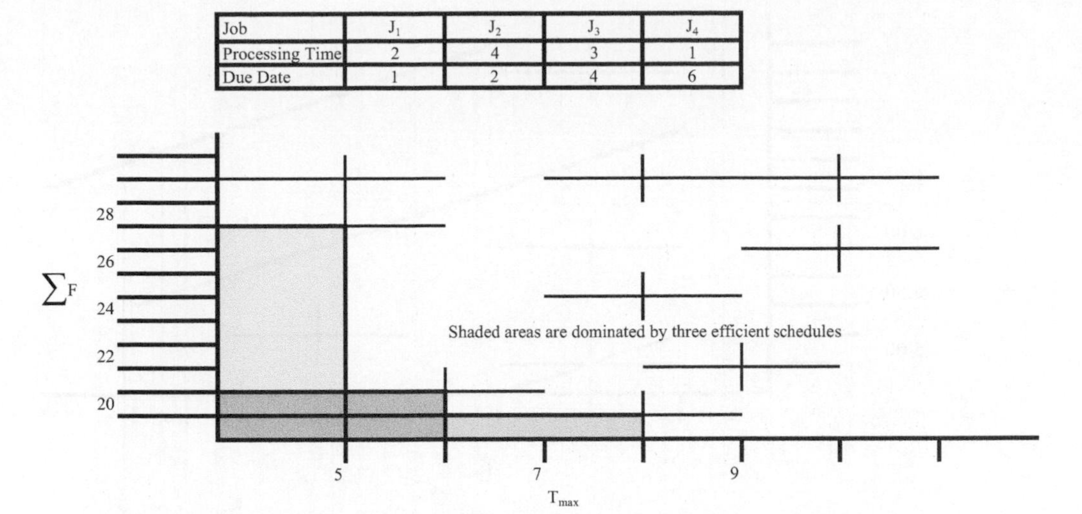

Figure 8.16 Data and Efficient Schedules for the Van Wassenhove and Gelders Example, Showing Some Pairs of T_{max} and $\overline{F}$ that are Obtained by Schedules of the Example. The Schedules in the Upper Right Hand Corner of the Shaded Areas are the Efficient Schedules

Step 3: $\Delta = 5 - 1 = 4$.

Step 4: $\Delta \geq 0$.

Step 2: We start our sequence with J_4 in last position, but after that we have no jobs with due dates greater or equal to $\tau = 9$.

Step 5: Stop.

In Figure 8.16 we have plotted some of the pairs (T_{max}, $\sum F$) that are obtainable by schedules in this example. The three schedules found above are clearly efficient, because no other schedules dominate them. Given the function f, the best schedule is (J_4, J_1, J_2, J_3) with f = 39.75.

Suppose the total cost function in this example is a different one with linear and positive coefficients: $f(T_{max}, \overline{F}) = 0.5T_{max} + 5\overline{F}$.

Note that this $f(T_{max}, \overline{F})$ is increasing in both its arguments, so as we have argued in the introduction, the minimal total cost function must be efficient. The three efficient schedules have the following total costs.

(J_4, J_1, J_3, J_2): $T_{max} = 8$. $\overline{F} = 20/4 \Rightarrow$ total cost $= 0.5 \times 8 + 5 \times 5.00 = 29.00$
(J_4, J_1, J_2, J_3): $T_{max} = 6$. $\overline{F} = 21/4 \Rightarrow$ total cost $= 0.5 \times 6 + 5 \times 5.25 = 29.25$
(J_1, J_2, J_3, J_4): $T_{max} = 5$. $\overline{F} = 27/4 \Rightarrow$ total cost $= 0.5 \times 5 + 5 \times 6.75 = 36.25$

The minimal total cost schedule then becomes (J_4, J_1, J_3, J_2). The choice of f is dependent on the user's preference among the two measures. There is an interesting and important geometric interpretation of this calculation that will be familiar

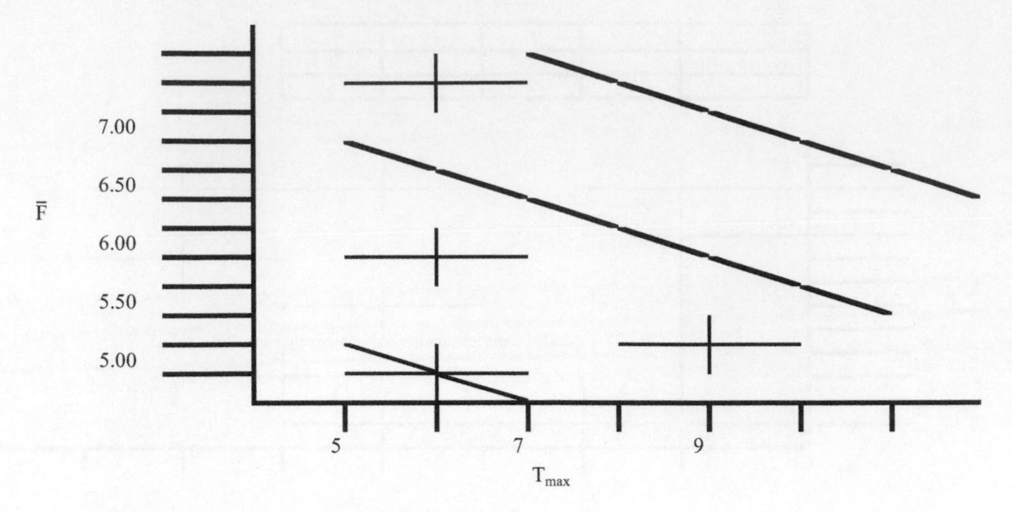

Figure 8.17 Parallel Lines Defined by f = c

to those who have studied mathematical programming. In Figure 8.17 we have plotted the family of lines given by $4T_{max} + 3\overline{F} = c$ for varying values of c. As c decreases, these lines move across the figure as shown. Moreover, each line joins points $(T_{max}, \overline{F})$ that are equal in terms of total cost. The point (6, 21) corresponding to schedule (J_4, J_1, J_2, J_3) lies on the total cost line $4T_{max} + 3\overline{F} = 39.25$, and it is clear that no line of lower total cost passes through an efficient schedule. Thus we see that this schedule achieves the minimum total cost.

9 ALGORITHMS FOR TWO-MACHINE PROBLEMS AND EXTENSIONS TO MULTIPLE MACHINES

Introduction

We now turn to problems with more than one machine, i.e., scheduling in the flow shop or general job shop. Our immediate concern is to study problems for which constructive algorithms exist. By a constructive algorithm we mean one that builds up an optimal solution from the data of the problem by following a simple set of rules, which exactly determine the processing order. In the single-machine case we encountered a variety of problems that could be solved constructively. Here we will not be so lucky; only a few with two or more machines are amenable to such analysis. The $n/2/F/F_{max}$ family of problems is the only one for which there exists a constructive algorithm applicable to all cases. For all other families the few constructive algorithms that exist apply only to special cases, usually cases in which the processing times are restricted in some way. In this chapter we consider the most important of these algorithms, namely those due to Johnson (1954).

Our assumption for this chapter is that $r_i = 0$ for all J_i, $i = 1, 2, \ldots, n$.

Some Results for Schedules in a Flow Shop

Because many operations can be described as flow shops with multiple machines, it would be ideal to be able to schedule these optimally. In a flow shop the technological constraints demand that the jobs pass through the machines in the same order; i.e., if J_1 must be processed on machine M_k before machine M_j. then the same is true for all jobs. Most production lines are arranged in this manner.

We would like to derive algorithms that can handle the largest possible family of flow shops. To this end we start with two theorems. We shall see that these imply that for some 2 and 3 machine flow shops we need only consider permutation schedules. A permutation schedule is one in which each machine processes the jobs in the same order; i.e., if on machine M_1 job J_j is processed before J_k, then the same is true for all machines. This is an important result as there are far fewer permutation schedules and they are also easier to implement because the person executing the schedule can use the same processing order on all machines. In flow shops there is a natural ordering of the machines, namely that given by the technological constraints as the processing order for each and every job. We assume that each job must be processed by M_1 before M_2 before M_j, etc. The technological

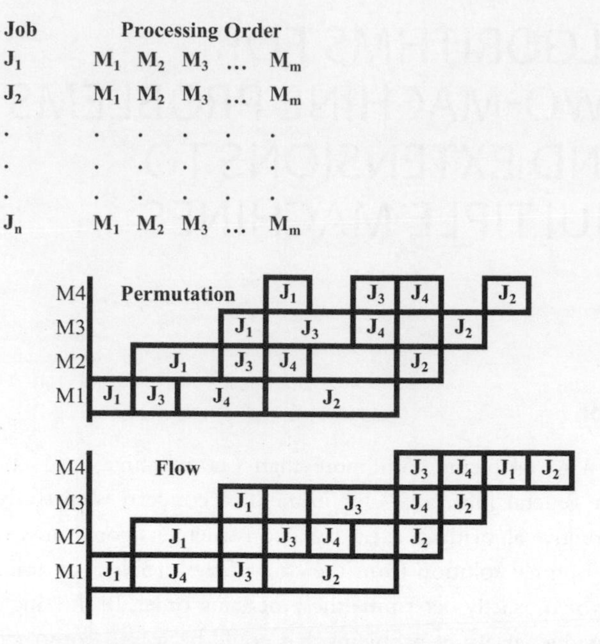

Figure 9.1 The Technological Constraints in a Flow Shop, Flow and Permutation Example

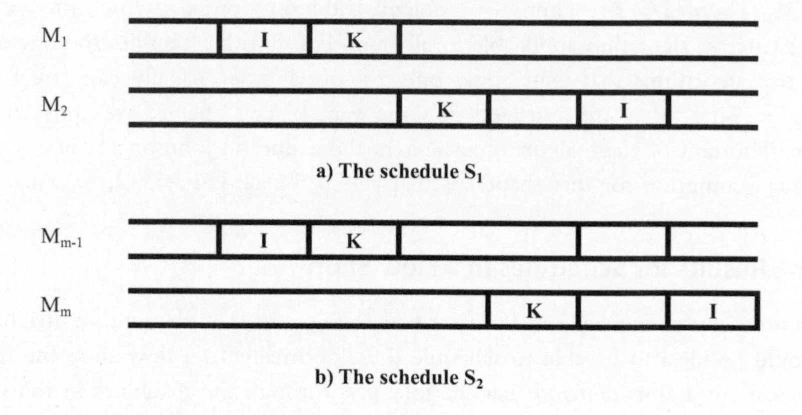

Figure 9.2 Exchanging Jobs for Theorems 1 and 2

constraints therefore have the form in Figure 9.1, which also has an example of a four-job, four-machine problem scheduled both as flow and permutation.

Theorem 9.1 For the n/m/F/B problem with B a regular measure of performance, it is sufficient to consider schedules in which the same processing sequence is given on the first two machines.

Proof

(i) If a schedule S_1 does not have the same order on both machines M_1 and M_2 there is a job I which directly precedes job K on M_1 but follows K perhaps with intervening jobs, on M_2 (see Figure 9.2(a)).

(ii) On M_1 we may reverse the order of I and K without increasing any starting time on M_2 Thus this interchange cannot increase the completion time of any job and, hence, neither can it increase a regular measure.
(iii) This process of exchanging jobs on M_1 may be repeated until a schedule S_1' is obtained with the same order on M_1 as on M_2. Clearly S_1' is no worse than S_1 under a regular measure.

Theorem 9.2 For the $n/m/F/F_{max}$ problem, there is no need to consider schedules with different processing orders on machines M_{m-1} and M_m.

Proof

(i) If a schedule S_2 does not have the same order on both machines M_{m-1} and M_m, there is a job I which directly precedes job K on M_{m-1} but follows K, perhaps with some intervening jobs, on M_m (see Figure 9.2(b)).
(ii) Suppose we reverse the processing order of I and K on M_m. Clearly this may change the flow times of individual jobs: some may increase; some may decrease.
(iii) However, in total the processing on M_m, can only be expedited, because I completes on M_{m-1} before K does. Thus F_{max} cannot be increased by the interchange.
(iv) This process of interchanging jobs on M_m may be repeated until a schedule S_2' is obtained with the same order on M_{m-1} as on M_m. The above shows that S_2' can be no worse than S_2 in terms of F_{max}.

The proofs of Theorems 9.1 and 9.2 are very similar, but by no means the same. The similarity is evident from Figures 9.2(a) and (b); these differ only in the labeling of the machines. The important distinction between the proofs lies in the machine chosen for the interchange of processing order. In the proof of Theorem 9.1 we reversed the order of I and K on M_1, the first of the two machines considered. We saw that doing so could improve and certainly not worsen each and every flow time. Ideally we should have liked to copy this interchange in the proof of Theorem 9.2, i.e., to have interchanged jobs on M_{m-1}. For then that theorem would have been applicable to problems of scheduling against any regular measure and not just F_{max}. However, further analysis shows that the effect of interchanging jobs on M_{m-1} is entirely unpredictable; what happens depends upon the times jobs complete on M_m. Thus we must make the interchange on M_m and so can only deduce a result in terms of F_{max}.

Due to Theorem 9.1 we need only consider permutation schedules for the n/2/F/B problem, when B is regular. Adding the result of Theorem 9.2, again we need only consider permutation schedules for the $n/3/F/F_{max}$ problem. These results have not been strengthened to more machines in the 60 or so years since Johnson.

Johnson's Algorithm for the $n/2/F/F_{max}$ Problem

Suppose we have an $n/2/F/F_{max}$ problem; i.e., we have to process n jobs through two machines, each job in the order M_1, M_2 so that the maximum flow time is minimized. Because all jobs have zero ready times, $F_{max} = C_{max}$. It is sensible to

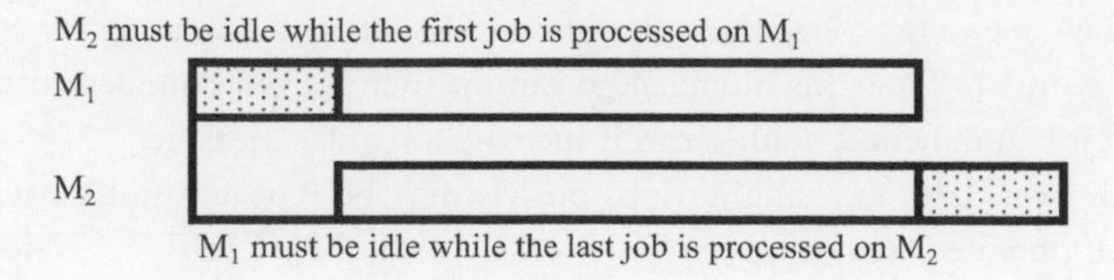

Figure 9.3 Two Machine n Job F_{max} Schematic

start the processing with the job with the shortest processing time on M_1. For then the processing on M_2 may start as soon as possible (see Figure 9.3).

Similarly it is just as sensible to finish the processing with the job that has the shortest processing time on M_2. For while this job is processing M_1 must be idle. Also we have just shown that we only need to consider permutation schedules for this problem. Putting these ideas together, it is reasonable to suggest that the optimal schedule is a permutation of $\{J_1, J_2, \ldots, J_n\}$ such that the earlier jobs in the processing sequence have short M_1 processing times, whereas the later jobs have short M_2 processing times. The schedule constructed by Johnson's algorithm has precisely this property.

We give Johnson's algorithm next and an example of its use. Then we prove that his method does find an optimal schedule. It will simplify our notation to let:

> $a_i = p_{i1}$ the processing time of the J_i on M_1; and $b_i = P_{i2}$ the processing time of the J_i on M_2;

The algorithm builds up the processing sequence by working in from the ends toward the middle. To do this we shall need two counters, namely k and i. k = 1 initially and increases 2, 3, 4 . . . as the first, second, third, fourth . . . positions in the processing sequence are filled. Similarly, i = n initially and decreases (n – 1), (n – 2), . . . as the nth, (n – l)th, (n – 2)th, . . . positions in the processing sequence are filled.

Algorithm 9.1 (Johnson)

Step 1: Set k = 1, i = n.
Step 2: Set the current list of unscheduled jobs = $\{J_1\ J_2, \cdot J_j, \ldots, J_n\}$.
Step 3: Find the smallest of all the a_j and b_j times for the currently unscheduled jobs.
Step 4: If the smallest time is for J_j on the first machine, i.e., a_j is smallest, then:

(i) Schedule J_j in kth position of processing sequence.
(ii) Delete J_j from current list of unscheduled jobs.
(iii) Increment k to k + 1.
(iv) Go to step 6.

Step 5: If the smallest time is for J_j on second machine, i.e., b_j is smallest, then:

(i) Schedule J_j in the ith position of processing sequence.
(ii) Delete J_j from current list of unscheduled jobs.
(iii) Reduce i to (i – 1).
(iv) Go to step 6.

Step 6: If there are any jobs still unscheduled, go to step 3. Otherwise, stop.

If the smallest time occurs for more than one job on two different machines in step 3, then pick J_j arbitrarily. If the smallest time occurs for more than one job on the same machine in step 3, then pick J_j with the longer processing time on the other machine. If these are also the same, then pick arbitrarily.

Example Schedule the $7/2/F/F_{max}$ problem with the following data (Figure 9.4). Applying the algorithm, the schedule builds up as follows:

The smallest processing time is 1 and occurs on both Jobs 4 and 5. Because job 4's processing time is on machine 1, it is placed first and because job 5's processing time is on machine 2, it is placed last.

Job 4 scheduled: 4 _ _ _ _ _ _
Job 5 scheduled: 4 _ _ _ _ _ 5

The next lowest processing time is 2 and it is for job 2 on machine 1. Therefore 2 is placed in the first open position in the front of the sequence.

Job 2 scheduled: 4 2 _ _ _ _ 5

We continue, using the rules of the algorithm

Job 1 scheduled: 4 2 _ _ _ 1 5
Job 3 scheduled: 4 2 _ _ 3 1 5
Job 6 scheduled: 4 2 6 _ 3 1 5
Job 7 scheduled: 4 2 6 7 3 1 5

Thus we sequence the jobs in the order (4, 2, 6, 7, 3, 1, 5).

It remains to actually calculate the maximum completion time by time tabling (Figure 9.5).

Processing Time on Machine		
Job	M_1	M_2
1	6	3
2	2	9
3	4	3
4	1	8
5	7	1
6	4	5
7	7	6

Figure 9.4 Data for the $n/2/F/F_{max}$ Example

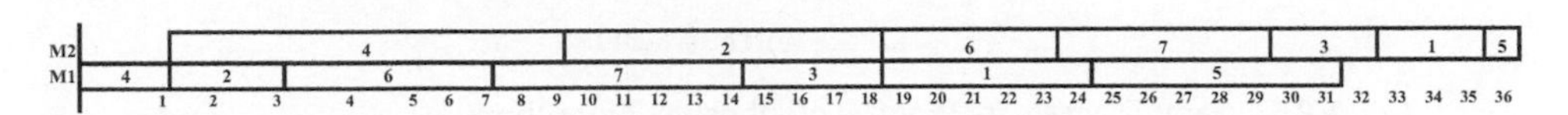

Figure 9.5 Gantt Chart for the $n/2/F/F_{max}$ Example

Our schedule has an F_{max} of 36. The Gantt chart also shows us why this sequence is optimal—both machines process continuously and M_2 starts as early as possible, given the processing times. When we explore all 7! = 5040 possible schedules, we find that 273 of them are optimal. Incidentally, the worst schedule has a completion time of 52, showing that it really pays to find the best schedule (an improvement of 31%!), because the initial sequence is arbitrary and can quite possibly be the worst. We now prove that Johnson's algorithm does produce an optimal schedule. We begin by showing that the first cycle of the algorithm chooses and positions a job optimally.

Theorem 9.3 For the n/2/ F/ F_{max} problem with $p_{i1} = a_i$ and $p_{i2} = b_i$ i = 1, 2, . . ., n:

I. If $a_k = \min\{a_1, a_2 . . ., a_n, b_1\ b_2, . . ., b_n\}$, there is an optimal schedule with J_k first in the processing sequence;
II. If $b_k = \min\{a_1, a_2 . . ., a_n, b_1\ b_2, . . ., b_n\}$, there is an optimal schedule with J_k last in the processing sequence.

Proof We prove (I.). Let J_k be such that $a_k = \min\{a_1, a_2 . . ., a_n, b_1\ b_2, . . ., b_n\}$. Let S be a schedule with J_k not first in the processing sequence. Let J_l be the job that is processed immediately before J_k. Then S has a Gantt diagram as given in Figure 9.6.

(i) Let Q be the time that M_1 is ready to start processing J_l under S. Let R be the time that M_2 finishes processing the job before J_l under S and is thus ready to process J_l. We consider the schedule S′ obtained simply by interchanging the jobs J_k and J_l in S. Our aim is to show that $C_k \geq C'_l$ where C_k is the completion time of J_k under S and C'_l is the completion time of J_l under S′. If this is true $F_{max} = C_{max} \geq C'_{max} = F'_{max}$ and we deduce that (i) holds by repeatedly interchanging J_k with the jobs before it until it is first to be processed.

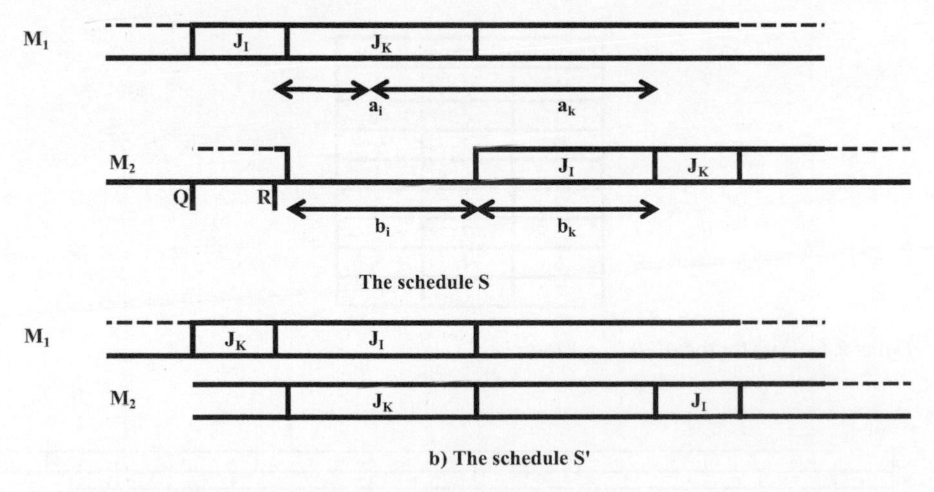

Figure 9.6 Gantt Diagram for Proof (I.) of Johnson's Algorithm

(ii) Under S J_l starts on M_2 at $\max\{R, Q + a_l\}$. We know, by assumption, that $a_k \leq b_l$. Thus J_l completes on M_2 after J_k completes on M_1 and so we have: $C_k = \max\{R,Q + a_l\} + b_l + b_k$.

(iii) Under S′, J_k starts on M_2 at $\max\{R, Q + a_k\}$ and J_l starts on M_2 either as soon as J_k finishes or as soon as M_1 finishes processing J_l, whichever is the sooner.

(iv) $C'_l = \max\{\max\{R, Q + a_k\} + b_k, Q + a_k + a_l\} + b_l$.

(v) $C'_l = \max\{\max\{R, Q + a_k\} + b_k + b_l, Q + a_k + a_l + b_l\}$.

(vi) $C_k = \max\{R, Q + a_l\} + b_l + b_k \geq \max\{R,Q + a_k\} + b_l + b_k$ because $a_l \geq a_k$ by definition.

(vii) $C_k = \max\{R, Q + a_l\} + b_l + b_k \geq Q + a_l + b_l + b_k$.

(viii) $C_k = \max \geq Q + a_l + b_l + a_k$ because $b_k \geq a_k$ by definition.

(ix) So C_k is no less than either of the terms in the outer max{ } of $C'_l = \max\{\max\{R, Q + a_k\} + b_k + b_l, Q + a_k + a_l + b_l\}$.

(x) Thus $C_k \geq C'_l$ as required.

The proof of (II.) is left for the reader. Now we deduce that Johnson's algorithm also positions the remaining (n – 1) jobs correctly.

Theorem 9.4 Algorithm 9.1 finds an optimal schedule for the $n/2/F/F_{max}$ problem.

Proof By Theorem 9.3 we know that the first cycle of the algorithm positions a job optimally. We now show that, if the first v cycles have positioned v jobs optimally the (v + 1)st cycle will also position a job optimally. Our result then follows by induction.

(i) Suppose v cycles have positioned v jobs optimally. Because of this optimality the (v + 1)st cycle need not reconsider any of the choices made in earlier cycles; it may simply choose one of the remaining (n – v) jobs and position it in one of the remaining (n – v) places in the processing sequence. What the cycle actually does is to choose and position a job according to the conditions of Theorem 9.3 applied to the remaining (n – v) unscheduled jobs alone. Look again at the proof of that theorem. The argument shows that $C_k \geq C'_l$ for the interchanged jobs. Moreover, only quantities directly relating to J_k and J_l enter the argument; the rest of the processing sequence is irrelevant.

(ii) It follows that, in placing the (v + 1)st job to minimize F_{max} for the remaining (n – v) jobs considered alone, the algorithm is minimizing F_{max} for all n jobs. Thus the (v + 1)st cycle positions a job optimally.

(iii) We know that the first cycle positions a job optimally. Setting v = 1, the above shows that the second cycle also positions a job optimally.

(iv) Next setting v = 2 we deduce that the third cycle does so, too.

(v) Continuing in this way, we deduce that all cycles position jobs optimally. Thus the algorithm constructs an optimal schedule.

Johnson's Algorithm for the $n/2/G/F_{max}$ Problem

Earlier we made the assumption that each job must be processed through all the machines. For this section we drop that assumption. Suppose that the set of n jobs $\{J_1, J_2, \ldots, J_n\}$ may be partitioned into four types of job as follows:

Type A: Those to be processed on machine M_1 only.
Type B: Those to be processed on machine M_2 only.
Type C: Those to be processed on both machines in the order M_1 then M_2.
Type D: Those to be processed on both machines in the order M_2 then M_1.

Then the construction of an optimal schedule follows:

Schedule the jobs of type A in any order to give the sequence S_A.
Schedule the jobs of type B in any order to give the sequence S_B.
Schedule the jobs of type C according to Johnson's algorithm for $n/2/F/F_{max}$ problem to give the sequence S_C.
Schedule the jobs of type D according to Johnson's algorithm for $n/2/F/F_{max}$ problems to give the sequence S_D. (Note: Here M_2 is the first machine and M_1 the second in the technological constraints.)

An optimal schedule then is:

Machine	Processing Order
M_1	(S_C, S_A, S_D)
M_2	(S_D, S_B, S_C)

To see that this schedule is optimal remember that time is wasted and hence F_{max} is increased, if either M_2 is kept idle waiting for jobs of type C to complete on M_1 or M_1 is kept idle waiting for jobs of type D to complete on M_2. This schedule minimizes such idle time.

Example Given $8/2/G/F_{max}$ problem with times and processing order given in Figure 9.7.

	Processing order and times			
Job	First Machine		Second Machine	
1	M_1	8	M_2	2
2	M_1	7	M_2	5
3	M_1	4	M_2	7
4	M_2	6	M_1	4
5	M_2	5	M_1	3
6	M_1	9		
7	M_2	1		
8	M_2	5		

Figure 9.7 Data for the $8/2/G/F_{max}$ Example

	Processing Sequence
Machine M_1	(3, 2, 1, 6, 4, 5)
Machine M_2	(4, 5, 7, 8. 3, 2, 1)

Figure 9.8 Solution to the 8/2/G/F_{max} Example

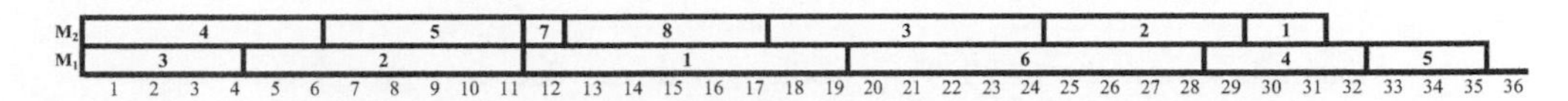

Figure 9.9 Gantt Chart for the n/2/G/F_{max} Example

To find an optimal schedule:

- Type A jobs—only job 6 is to be processed on M_1 alone.
- Type B jobs—jobs 7 and 8 require M_2 alone. Select arbitrary order (7, 8).
- Type C jobs—jobs 1, 2, and 3 require M_1 first and then M_2. Johnson's algorithm for this 3/2/F/F_{max} problem gives the sequence (3, 2, 1).

Type D jobs—jobs 4, 5 require M_2 first and then M_1. Johnson's algorithm for this 2/2/F/F_{max} problem gives the sequence (4, 5) (remember M_1 is now the second machine).

The optimal sequence for the example as generated is shown in Figure 9.8.

The resulting Gantt diagram is given in Figure 9.9. From this we see that F_{max} = 35 for an optimal schedule. It is also worth pointing out that there are 3,628,800 schedules (6! × 7!) without Johnson's algorithm.

A Special Case of the n/3/F/F_{max} Problem

Johnson's algorithm for the n/2/F/F_{max} problem may be extended to a special case of the n/3/F/F_{max} problem. We need the condition that:

$$\text{either } \min_{i=1}^{n}\{p_{i1}\} \geq \max_{i=1}^{n}\{p_{i2}\}$$
$$\text{or } \min_{i=1}^{n}\{p_{i3}\} \geq \max_{i=1}^{n}\{p_{i2}\} \qquad \text{Equ. 9.1}$$

That is, the maximum processing time on the second machine is no greater than the minimum time on either the first or the third. If Equation 9.1 holds, an optimal schedule for the problem may be found by letting

$$a_i = p_{i1} + p_{i2}$$
$$b_i = p_{i2} + p_{i3}$$

and scheduling the jobs as if they are to be processed on two machines only, but with the processing time of each job being a_i and b_i on the first and second machines respectively. To obtain the actual F_{max}, one must of course generate the Gantt diagram for the original problem.

Remember we have shown that we need only consider permutation schedules for the $n/2/F/F_{max}$ problem and note that Johnson's algorithm applied to the $n/2/F/F_{max}$ problem with the constructed a_i and b_i times does produce such a schedule. We could also construct an example, which shows that the condition of Equation 9.1 is necessary for this method to find an optimal schedule.

Example Given the $6/3/F/F_{max}$ problem with the data in Figure 9.10. First we check that Equation 9.1 holds for this problem. Here we have

$$\min_{i=1}^{6}\{p_{i1}\} = 3;\ \max_{i=1}^{6}\{p_{i2}\} = 3;\ \min_{i=1}^{6}\{p_{i3}\} = 2$$

Thus we have $\min_{i=1}^{6}\{p_{i1}\} = 3 \geq 3 = \max_{i=1}^{6}\{p_{i2}\}$ and Equation 9.1 holds. This is an either–or condition; we do not need both inequalities to hold. The constructed a_i and b_i times are also in Figure 9.10. Applying Algorithm 9.1 gives the sequence of placements:

Job 6 scheduled: _ _ _ _ _ 6 (The minimum 3 occurred on the "second" machine).

Job 3 scheduled: 3 _ _ _ 1 6 (The minimum 4 occurred on both machines, so we can schedule them in either order, or simultaneously) until we arrive at (3, 4, 5, 2, 1, 6), with the Gantt chart in Figure 9.11. Complete enumeration shows that there are only two optimal schedules. The worst schedule has F_{max} = 40.

	Actual Processing Times			Constructed Processing Times	
Job	M_1	M_2	M_3	1st Machine	2nd Machine
1	4	1	3	5	4
2	6	2	4	8	6
3	3	1	4	4	5
4	5	3	6	8	9
5	8	2	6	10	8
6	4	1	2	5	3

Figure 9.10 Data for the $6/3/F/F_{max}$ Example

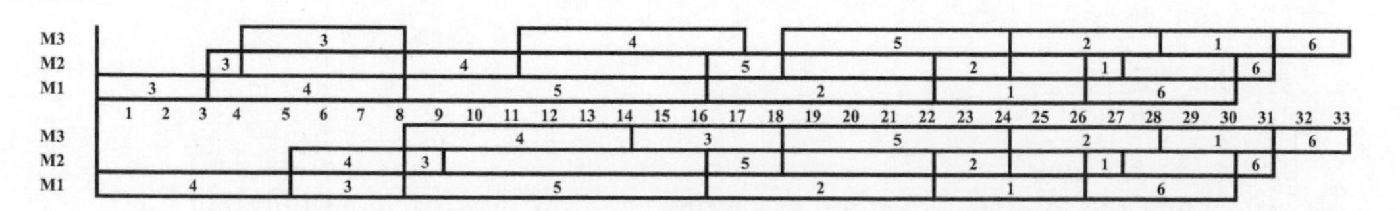

Figure 9.11 Gantt Chart for the Two Optimal Schedules for the $n/3/F/F_{max}$ Example

You can see that M_2 does not delay the processing on M_3 in any way. One can ask what happens if the conditions of Equation 9.1 are not met? We can no longer claim to get an optimal schedule, but do tend to get very good schedules as we will see in later chapters. The implications of these conditions are considerable. Whenever we are to process flow jobs through a number of machines, we are well advised to keep processing times as equal to each other as possible, thus diminishing the need to schedule in detail. Also, by keeping processing times on interior machines lower, we can concentrate on scheduling the outer machines.

10 IMPLICIT ENUMERATIONS

In this chapter we consider two enumeration methods: dynamic programming and branch and bound. These list, or enumerate, all possible schedules and then eliminate the non-optimal schedules from the list, leaving those that are optimal. Both of these methods are applicable to many types of problems, not just scheduling.

Dynamic Programming Approaches

Dynamic programming originates from Bellman (1957). White (1969) provides an introductory treatment and an extensive review of its applications; also most general operations research textbooks discuss it briefly. The method applies to any problem that can be broken down into a sequence of nested problems, the solution of one being derived in a straightforward fashion from that of the preceding problem.

The Approach of Held and Karp

In 1962 Held and Karp followed the lead of Bellman and applied dynamic programming ideas to sequencing problems. Their method applies to single-machine problems where the performance measure takes the form: $\sum_{i=1}^{n} \gamma_i \left(C_i\right)$.

Here the γ_i (C_i) are assumed to be non-decreasing functions of the completion times. It follows that this performance measure is regular. Particularly note that $\bar{C}$, $\bar{F}$ and $\bar{T}$ take this form. (Set γ_I (C_i) equal to C_i/n, $(C_i - r_i)/n$ and $\max\{C_i - d_i, 0\}/n$, respectively.) Thus we may use Held and Karp's method to solve $n/1//\bar{C}$, $n/1//\bar{F}$, and $n/1//\bar{T}$ problems. It would be foolish to use it on the first two problems; we know that an SPT schedule solves them. But we have been unable to solve $n/1//\bar{T}$ problem until now. The approach may also be extended to solve problems with performance measures of the form: $\left\{max_{i=1}^{n}\gamma_i\left(C_i\right)\right\}$. For this dynamic programming is not the best method of solution; Lawler's algorithm is. Held and Karp's approach is based upon an observation about the structure of an optimal schedule. This says that in an optimal schedule the first K jobs (any K = 1, 2, . . ., n) must form an optimal schedule for the reduced problem based on just these K jobs alone. Suppose that $(J_{i(1)}, J_{i(2)}, \ldots, J_{i(n)})$ is an optimal schedule for the full problem. Then for any K= 1, 2, . . . n we may decompose the performance measure as:

$$\sum_{k=1}^{n} \gamma_{i(k)}\left(C_{i(k)}\right) = \sum_{k=1}^{A} \gamma_{i(k)}\left(C_{i(k)}\right) + \sum_{k=K+1}^{B} \gamma_{i(k)}\left(C_{i(k)}\right) = \mathrm{A} + \mathrm{B} \qquad \text{Equ. 10.1}$$

Consider the set of jobs $(J_{i(1)}, J_{i(2)}, \ldots, J_{i(k)})$. Suppose that we face the task of scheduling just these K jobs and not all n. Then surely we cannot improve upon the sequence $(J_{i(1)}, J_{i(2)}, \ldots, J_{i(k)})$. If we could, we would be able to reduce the term A in equation 10.1. Thus we could construct a new sequence for the full problem by using the improved sequence for the first K jobs and leaving the remaining (n – K) in their original order. The total cost of this new sequence would be the sum of a quantity strictly less than A plus the original term B. But this contradicts our basic assumption that $(J_{i(1)}, J_{i(2)}, \ldots, J_{i(K)})$ is an optimal sequence for the full problem. Thus $(J_{i(1)}, J_{i(2)}, \ldots, J_{i(K)})$ must be optimal for the reduced problem.

We now introduce some notation. When we write $\{J_1, J_2, \ldots, J_n\}$ we mean the set of jobs listed between the brackets; there is no implication of a particular processing order. However, when we use parentheses, namely $(J_1, J_2, \ldots, J_n)$, we imply an order: namely, J_1 first, J_2 second, . . ., J_n last. The former notation refers to a set, the latter to a sequence. If Q is any set of jobs containing the job J, then Q – {J} is the set of jobs obtained by deleting J from Q. Also for the set Q we define C_Q by

$$C_Q = \sum_{J_i \text{ in } Q} p_i \qquad \text{Equ. 10.2}$$

Next we define Γ(Q) to be the minimum cost obtained by scheduling the jobs in Q optimally. If Q contains a single job, say Q = $\{Q_i\}$, then,

$$\Gamma(Q) = \Gamma(\{J_i\}) = \gamma_i(p_i) \qquad \text{Equ. 10.3}$$

because there is only one way to schedule one job and that job must complete at $C_i = p_i$. Suppose now that Q contains K > 1 jobs. Then, remembering that the last job must complete at C_Q and that in any optimal sequence the first (K – 1) jobs are scheduled optimally for a reduced problem involving just those, we have

$$\Gamma(Q) = \min_{J_i \text{ in } Q} \{\Gamma(Q - \{J_i\}) + \gamma_i(C_Q)\} \qquad \text{Equ. 10.4}$$

To find the minimum cost of scheduling Q we consider each job in turn and ask how much it would cost to schedule that job last. We find our answer by taking the minimum of all these possibilities. Figure 10.2 defines Γ(Q) for all Q containing a single job. Using these values and Equation 10.4 we can find Γ(Q) for all Q containing just two jobs; then for all Q containing just three jobs; and so on until we eventually find $\Gamma(\{J_1, J_2, \ldots, J_n\})$. In doing so we also find an optimal schedule. It is easiest to see all this with an example.

Example Solve the $4/1//\overline{T}$ problem with the data in Figure 10.1.
Here $\gamma_i(C_i) = \max\{C_i - d_i, 0\}$ Note that $\overline{T}$ and sum of T are equivalent in solving an optimization problem and we save the many divisions by n. First, we calculate

Job J_i	J_1	J_2	J_3	J_4
Processing time p_i	8	6	10	7
Due date d_i	14	9	16	16

Figure 10.1 Parameters for the Dynamic Programming Example

Q	$\{J_1\}$	$\{J_2\}$	$\{J_3\}$	$\{J_4\}$
$p_i - d_i$	-6	-3	-6	-9
$\Gamma(Q)$	0	0	0	0

Figure 10.2 Penalties for Single Job Sets

Q	**$\{J_1, J_2\}$**		**$\{J_1, J_3\}$**		**$\{J_1, J_4\}$**		**$\{J_2, J_3\}$**		**$\{J_2, J_4\}$**		**$\{J_3, J_4\}$**	
C_Q	**14**		**18**		**15**		**16**		**13**		**17**	
J_i last job in the sequence	**J_1**	**J_2**	**J_1**	**J_3**	**J_1**	**J_4**	**J_2**	**J_3**	**J_2**	**J_4**	**J_3**	**J_4**
$\gamma_i(C_Q)$	**0**	**5**	**4**	**2**	**1**	**0**	**7**	**0**	**4**	**0**	**1**	**1**
$\Gamma(Q-\{J_i\})$	**0**	**0**	**0**	**0**	**0**	**0**	**0**	**0**	**0**	**0**	**0**	**0**
Sum	**0**	**5**	**4**	**2**	**1**	**0**	**7**	**0**	**4**	**0**	**1**	**1**
Minimum $\Gamma(Q)$	*			*		*		*		*	*	

Figure 10.3 The Calculations of $\Gamma(Q)$ for Each Two-Job Set

$\Gamma(Q)$ for the four sets that only have a single member. Using $\Gamma(\{J_1\}) = \max\{0, C_1 - d_1\} = \max\{0, p_1 - d_1\}$ we generate Figure 10.2.

Next we calculate $\Gamma(Q)$ for the six sets containing just two jobs. For instance, if $Q = \{J_1, J_2\}$, we have $C_Q = p_1 + p_2 = 14$, and by Equation 10.4,

$$\Gamma(Q) = \min\{\Gamma(\{J_1\}) + \gamma_2(14), \Gamma(\{J_2\}) + \gamma_1(14)\} = \min\{0 + 3, 0 + 0\} = 0$$

The calculations for this and the five other two-member subsets are laid out in Figure 10.3.

There are two possibilities for each Q corresponding to which job is scheduled last: the columns are divided accordingly. We mark the column corresponding to the minimum in Figure 10.3 by an asterisk, choosing arbitrarily whenever there is a tie. Then we calculate $\Gamma(Q)$ for the four sets containing three jobs. For instance, if $Q = \{J_1, J_2, J_3\}$, we have $C_Q = p_1 + p_2 + p_3 = 24$ and by Equation 10.4,

$$\Gamma(Q) = \min\{\Gamma(\{J_1, J_2\}) + \gamma_3(24), \Gamma(\{J_1, J_3\}) + \gamma_2(24), \gamma_3(\{J_2, J_3\}) + \gamma_1(24)\}$$
$$= \min\{0 + 8, 2 + 15, 0 + 10\} = 8.$$

Q	$\{J_1, J_2, J_3\}$			$\{J_1, J_2, J_4\}$			$\{J_1, J_3, J_4\}$			$\{J_2, J_3, J_4\}$		
C_Q	24			21			25			23		
J_i last job in the sequence	J_1	J_2	J_3	J_1	J_2	J_4	J_1	J_3	J_4	J_2	J_3	J_4
$\gamma_i(C_Q)$	10	15	8	7	12	5	11	9	9	14	7	7
$\Gamma(Q-\{J_i\})$	0	2	0	0	0	0	1	0	2	1	0	0
Sum	10	17	8	7	12	5	12	9	11	15	7	7
Minimum $\Gamma(Q)$			*			*		*			*	

Figure 10.4 The Calculations of Γ(Q) for Each Three-Job Set

Q	$\{J_1, J_2, J_3, J_4\}$			
C_Q	31			
J_i last job in the sequence	J_1	J_2	J_3	J_4
$\gamma_i(C_Q)$	17	22	15	15
$\Gamma(Q-\{J_i\})$	7	9	5	8
Sum	24	31	20	23
Minimum $\Gamma(Q)$			*	

Figure 10.5 The Calculations of Γ(Q) for the Entire Set of Four Jobs

The calculations for this and the other three-member sets are laid out in Figure 10.4. The construction of this parallels that of Figure 10.3 except for each Q there are now three possible last jobs and so each column subdivides into three.

Finally we calculate Γ(Q) for the set containing all four jobs, namely Q = $\{J_1, J_2, J_3, J_4\}$. Here we have $C_Q = p_1 + p_2 + p_3 + p_4 = 31$. We calculate Γ(Q) in Figure 10.5. The construction of this exactly parallels the two previous figures.

We have found that scheduling the four jobs optimally gives a minimum sum of tardiness of 20. To find the optimal schedule we trace our way back through the figures. The asterisk in the third column of the Figure 10.5 tells us that the minimum sum of tardiness is obtained when J_3 is scheduled last. This means that $\{J_1, J_2, J_4\}$ must be the first three jobs processed. We look at Figure 10.4 and find that of these J_4 should be scheduled last. Thus we know that $\{J_1, J_2\}$ are the first two jobs processed. Looking at Figure 10.3 we find that J_1 should be processed last. This leaves J_2 as the job first processed. So an optimal schedule is (J_2, J_1, J_4, J_3).

Dynamic: Programming Subject to Precedence Constraints

We have earlier made the statement that the introduction of precedence constraints into a problem far from complicating matters, can actually make it easier to solve (see Chapter 8). Let's explore this idea with our current topic of using dynamic programming to solve the $\overline{T}$ problem.

Example Consider the $5/1//\overline{T}$ problem with the data shown in Figure 10.6 and the precedence constraints that demand that J_1 is completed before either J_2 or J_3

Job	J_1	J_2	J_3	J_4	J_5
Processing time p_i	4	2	6	3	5
Due date d_i	7	9	9	6	12

Figure 10.6 Data for the Precedence Example Problem

Q	$\{J_1\}$	$\{J_4\}$
p_i - d_i	-3	-3
Γ(Q)	0	0

Figure 10.7 Single Sets of Jobs in the Precedence Example

Q	$\{J_1, J_2\}$		$\{J_1, J_3\}$		$\{J_1, J_4\}$		$\{J_4, J_5\}$	
C_Q	6		10		7		8	
J_i last job in sequence	J_1	J_2	J_1	J_3	J_1	J_4	J_4	J_5
$\gamma_i(C_Q)$	Impossible	0	Impossible	1	0	1	Impossible	0
$\Gamma(Q-\{J_i\})$		0		0	0	0		0
Sum		0		1	0	1		0
Minimum		*		*	*	*		*
Γ(Q)	0		1		0		0	

Figure 10.8 Feasible Sets of Two Jobs for the Precedence Problem

is started and that J_4 is completed before J_5 is started. There is no requirement that, say, J_4 immediately precedes J_5; other jobs may intervene in all cases.

So let us begin to calculate Γ(Q) for each possible set of jobs. As in the previous example we begin with the single job sets. We see right away that there is no point in considering the three sets $\{J_2\}$, $\{J_3\}$ and $\{J_5\}$ because none of these can be the first job in a schedule. Thus our first figure has two columns instead of five (Figure 10.7).

Next we consider the two-job sets. Given that there are five jobs in all, there should be five items taken two at a time, that is, 10 sets. However, only four of these are compatible with the precedence constraints. For instance, $\{J_1, J_5\}$ cannot be the first two jobs in any schedule because J_4 must precede J_5. (You may want to write all the sets to see why this is true). Moreover, when we consider those two-job sets that are possible, we find that we have no choice about which to sequence last in three of the four cases; the precedence constraints dictate the order. Thus, in Figure 10.8, instead of 10 columns each subdividing into two, we have four columns, only one of which subdivides.

When we continue we find similar reductions in the number of calculations that we have to perform. There are also 10 sets of three jobs, but six are eliminated

Q	{J₁, J₂, J₃}			{J₁, J₂, J₄}			{J₁, J₃, J₄}			{J₁, J₄, J₅}		
C_Q	12			9			13			12		
J_i last; job in sequence	J_1	J_2	J_{13}	J_1	J_2	J_3	J_1	J_3	J_4	J_1	J_4	J_5
$\gamma_i(C_Q)$	Impossible	3	3	Impossible	0	3	Impossible	4	7	5	Impossible	0
$\Gamma(Q-\{J_i\})$		1	0		0	0		0	1	0		0
Sum		4	3		0	3		4	8	5		0
Minimum			*		*			*				*
$\Gamma(Q)$	3			0			4			0		

Figure 10.9 Feasible Sets of Three Jobs for the Precedence Problems

Q	{J₁, J₂, J₃, J₄}				{J₁, J₂, J₄, J₅}				{J₁, J₃, J₄, J₅}			
C_Q	15				14				18			
J_i last; job in sequence	J_1	J_2	J_3	J_4	J_1	J_2	J_4	J_5	J_1	J_3	J_4	J_5
$\gamma_i(C_Q)$	Impossible	6	6	9	Impossible	5	Impossible	2	Impossible	9	Impossible	6
$\Gamma(Q-\{J_i\})$		4	0	3		0		0		0		4
Sum		10	6	12		5		2		9		10
Minimum			*									
$\Gamma(Q)$	6				2				9			

Figure 10.10 Feasible Sets of Four Jobs for the Precedence Problems

Q	{J₁, J₂, J₃, J₄, J₅}				
C_Q	20				
J_i last job in sequence	J_1	J_2	J_3	J_4	J_5
$\gamma_i(C_Q)$	Impossible	11	11	Impossible	8
$\Gamma(Q-\{J_i\})$		9	2		6
Sum		20	13		14
Minimum					
$\Gamma(Q)$	13				

Figure 10.11 Set of Five Jobs for the Precedence Problems

by the precedence constraints as shown in Figure 10.9. For the sets of four, we find that of the five candidates only three are feasible (Figure 10.10). We can now proceed to the one set of five (Figure 10.11).

Thus stepping back through the figures picking out the appropriate asterisks, we find that the optimal schedule is (J_4, J_1, J_2, J_5, J_3). The mean tardiness for this schedule is 13/5. The solution of this $5/1//\overline{T}$ problem required no more effort than our solution of the $4/1//\overline{T}$ problem. Yet, had there been no precedence constraints, it would have required much more. It has been shown (French, 1982) that it requires about $6n2^{n-1} + 3(2^n - 1)$ operations to solve a dynamic program for $n/1//\sum\gamma_i$, where n is the number of jobs. Using this expression we find that the solution of a five-job problem without precedence constraints requires about two and a half times as much work as that of a four-job one. We see that the introduction of precedence constraints makes our solution of $n/1//\sum\gamma_i$ problems by dynamic programming easier. Exactly how much easier depends on the particular precedence constraints, but clearly a substantial reduction in effort is possible. Many practical problems involve precedence constraints, so precedence constraints can be quite helpful.

As we solved the previous example, we needed to identify each of the subsequences specifically to determine which ones needed to be evaluated. We can avoid having to identify these if one works from the set of five. The precedence constraints eliminate jobs one and four from being last. Therefore, from the sets of four we only need 1345, 1245, and 1234. Again using the precedence constraints, for these we only need 145, 134, 124, and 123. Next we only need 45, 14, 13, and 12. Finally, we only need 1 and 4 (just as we did when we started with the sets of one). So we would need exactly the same number of calculations, but did not have to identify each of the possible sequences. Of course we still have to work our way back to the sets of one before we can fill in all the numbers. There is yet another way to make our solution of problems without precedence constraints easier. Consider the following theorem.

Theorem 10.1 In an $n/1//\bar{T}$ problem if two jobs J_j and J_k, are such that (i) $p_i \leq p_k$ and (ii) $d_i \leq d_k$ then there exists an optimal schedule in which J_i precedes J_k. Note that this does not imply that all optimal schedules will have such a relationship.

This theorem is typical of a class of results called dominance conditions or elimination criteria. These take the form: if J_i is related to J_k in a particular way, then there exists an optimal schedule in which J_j precedes J_k. Suppose that we have an $n/1//\bar{T}$ problem and suppose that we find a dominance condition applies to the pair of jobs J_j and J_k. Then we know that an optimal schedule exists in which J_i precedes J_k. Hence we may introduce the precedence constraint that J_i should precede J_k and be sure that a solution to the modified problem also solves the original one. The introduction of the precedence constraint makes the problem easier to solve by dynamic programming. If we examine enough pairs of jobs and find enough dominance conditions, we might substantially reduce a large problem and make it more easily solvable by dynamic programming. Because precedence conditions that are not based on dominance are imposed on a problem external to the parameters of a problem, such precedence constraints may invalidate some or all of the usefulness of the indicated dominance conditions. So let's return to our original example of the $n/1//\bar{T}$ problem and apply Theorem 10.1. From Figure 10.1 we can deduce the dominance condition that 1 must precede 4. The reduced problem is shown in Figure 10.12.

Empirical investigations have shown that this use of dominance conditions into a problem can lead to very fast dynamic programming algorithms for solving single-machine problems (Baker and Schrage, 1978; Schrage and Baker, 1978; Van Wassenhove and Gelders, 1980).

Q	{J$_1$}	{J$_2$}	{J$_3$}
p$_i$ - d$_i$	-6	-3	-6
Γ(Q)	0	0	0

Q	**{J$_1$, J$_2$}**		**{J$_1$, J$_3$}**		**{J$_1$, J$_4$}**		**{J$_2$, J$_3$}**	
C$_Q$	**14**		**18**		**15**		**16**	
J$_i$ last job in the sequence	**J$_1$**	**J$_2$**	**J$_1$**	**J$_3$**	**J$_1$**	**J$_4$**	**J$_2$**	**J$_3$**
γ$_i$(C$_Q$)	**0**	**3**	**4**	**2**	**impossible**	**0**	**7**	**0**
Γ(Q-{J$_i$})	**0**	**0**	**0**	**0**		**0**	**0**	**0**
Sum	**0**	**3**	**4**	**2**		**0**	**7**	**0**
Minimum Γ(Q)	*			*		*		*

Q	**{J$_1$, J$_2$, J$_3$}**			**{J$_1$, J$_2$, J$_4$}**			**{J$_1$, J$_3$, J$_4$}**		
C$_Q$	**24**			**21**			**25**		
J$_i$ last job in the sequence	**J$_1$**	**J$_2$**	**J$_3$**	**J$_1$**	**J$_2$**	**J$_4$**	**J$_1$**	**J$_3$**	**J$_4$**
γ$_i$(C$_Q$)	**10**	**15**	**8**	**impossible**	**12**	**5**	**impossible**	**9**	**9**
Γ(Q-{J$_i$})	**0**	**2**	**0**		**0**	**0**		**0**	**2**
Sum	**10**	**17**	**8**		**12**	**5**		**9**	**11**
Minimum Γ(Q)			*			*		*	

Q	**{J$_1$, J$_2$, J$_3$, J$_4$}**			
C$_Q$	**31**			
J$_i$ last job in the sequence	**J$_1$**	**J$_2$**	**J$_3$**	**J$_4$**
γ$_i$(C$_Q$)	**impossible**	**22**	**15**	**15**
Γ(Q-{J$_i$})		**9**	**5**	**8**
Sum		**31**	**20**	**23**
Minimum Γ(Q)			*	

Figure 10.12 Dominance Applied to the n/1//$\overline{T}$ Problem

Branch and Bound

Besides heuristic methods, branch and bound is probably the solution technique most widely used in scheduling. Like dynamic programming, it is an enumeration technique and like dynamic programming, it is an approach to optimization that applies to a much larger class of problems than just those in scheduling. Before we look at branch and bound, let's take another look at the logical structure of dynamic programming. The two are related, and appreciating that relationship will further the understanding of dynamic programming and serve as a useful introduction to the study of branch and bound. Although our examples will be based upon fairly simple classes of problems, there are no specific limitations to the branch and bound method.

Dynamic Programming and Its Elimination Tree

Recall the 4/1//$\overline{T}$ example earlier in this chapter and considering how dynamic programming eliminates the non-optimal schedules. A schedule for this problem corresponds to an assignment of a different job to each of the four positions in the processing sequence. We may generate all 4! = 24 possible schedules through the hierarchical, branching structure shown in Figure 10.13. For now ignore the distinction between solid and dotted lines. We start the generation of a schedule with no job sequenced and indicate this by the point or node XXXX. An 'X' in the processing sequence indicates that so far no job has been assigned that position.

We begin by assigning a job to the first position in the sequence. Each circle represents a partial or complete schedule and is referred to as a node. We move from node XXXX to one of the four mutually exclusive and exhaustive possibilities: lXXX, 2XXX, 3XXX, and 4XXX. Next we assign the second job in the sequence, so branching from each of these four nodes to three possibilities. 1XXX branches to the nodes 12XX, 13XX, and 14XX; 2XXX branches to the nodes 21XX, 23XX, and 24XX; and so on. Finally we subdivide our possibilities by assigning the job to be processed third. Because there are only four jobs, the last job is also positioned. For example, 12XX branches to the nodes 1234 and 1243. The branching structure in Figure 10.13 is called an elimination tree because it is used to determine the elimination of non-optimal schedules. Look at the calculation of the mean tardiness of any schedule, say 2413. We start by taking the job scheduled first, calculate its completion time, its tardiness and, consequently, its contribution to $\bar{T}$. This calculation is represented by a move from node XXXX to node 2XXX. The calculation of the contribution to $\bar{T}$ of the job scheduled second is similarly represented by a move from node 2XXX to node 24XX. Third and fourth we calculate the tardiness of jobs 1 and 3. We represent these calculations by moves from 24XX to 241X and then from 241X to 2413. However, nodes 241X and 2413 are consolidated in our diagram so this move is not fully indicated. The calculation of $\bar{T}$ for any schedule may be represented by moves along branches of the tree starting at XXXX and traveling to the appropriate final node. Each time a node is encountered, a further contribution to $\bar{T}$ is evaluated.

With this tree representation we may discuss the structure of different solution methods. For example, complete enumeration systematically starts from node XXXX 24 times and travels to each and every final node. In doing so it calculates all the $\bar{T}$s. This requires many computations.

Dynamic programming again starts at XXXX; but here there are four simultaneous moves to nodes 1XXX, 2XXX, 3XXX, and 4XXX. These correspond to the calculations in Figure 10.2, which evaluate $\Gamma(Q)$ for the single-member sets. Next consider the calculation of $\Gamma(Q)$ for the two-member sets. For instance, look at $Q = \{J_1, J_2\}$ in Figure 10.3. From our calculations we determine that job 1 should be scheduled last. In terms of the elimination tree, moving from 2XXX to 21XX is better than moving from 1XXX to 12XX. This we indicate by a solid line for the former and a dotted for the latter. We also know that we don't have to consider nodes that lie beyond 12XX in our tree. We use dotted lines to join node 12XX to both 1234 and 1243. We have eliminated a pair of final nodes from further consideration. Our calculations for each of the other five two-member sets similarly eliminate five other pairs of final nodes. Of the 24 possible schedules we have already eliminated 12. Turning to the calculation of $\Gamma(Q)$ for three-member sets Q, we see that further schedules are eliminated. Consider $Q = \{J_1, J_2, J_3\}$. Here

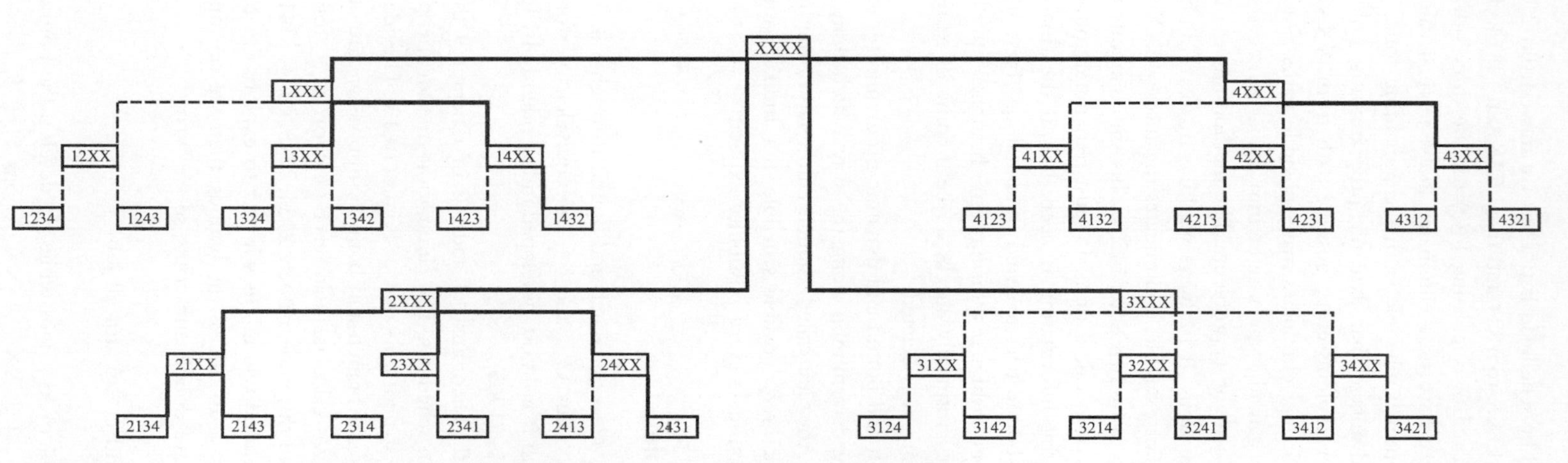

Figure 10.13 The Elimination Tree for the $4/1//\overline{T}$ Problem of the Dynamic Programming Example

we learn that J_1 should be scheduled last. Thus we move from 21XX to 2134 and eliminate moves from 13XX to 1324 and from 23XX to 2314. Three other pairs of schedules are eliminated in calculating $\Gamma(Q)$ for the three other sets. So of the 24 possible schedules we have now eliminated 20. The final calculation of $\Gamma(\{J_1, J_2, J_3, J_4\})$ selects an optimal schedule from the remaining four.

You can see the advantage that dynamic programming holds over complete enumeration. The latter explores every possible path from XXXX to a complete schedule, whereas the former eliminates many possible paths along the way.

From the above we gained some insight into the advantages of dynamic programming, but also to some of its problems. The method moves out simultaneously from XXXX in all directions. It is always working in many different parts of the tree at the same time. For example, it eliminates the move 43XX to 4312 by consideration of the move from 14XX to 1432. This simultaneous working requires many references to previous calculations. It must remember much information at each stage and all of this information is needed until the final selection of the optimum. Branch and bound tries to avoid these failings. First, it moves out from XXXX in a very uneven fashion, exploring some branches fully before looking at others at all. Second, it continually checks to see if some of stored information is still needed, and if it isn't, it is deleted.

Because both branch and bound and dynamic programming explore the elimination tree intelligently, determining along the way which branches do not need to be fully investigated, they are called implicit enumeration methods. It is implicit in their logic to check every possible schedule, but unlike complete or explicit enumeration neither considers every possibility explicitly.

A Flow Shop Example

We introduce the basic ideas of branch and bound through a $4/3/P/C_{max}$ example with zero ready times. Thus C_{max} and F_{max} are equivalent performance measures. The formulation is taken from those developed independently by Ignall and Schrage (1965) and by Lomnicki (1965).

We have shown with Theorems 9.1 and 9.2 in Chapter 9 that we only need to consider permutation schedules. The elimination tree for this problem will have an identical branching structure to that of Figure 10.13. The idea of bounding is central to the technique of branch and bound. Suppose we are at node 2XXX in the elimination tree. We calculate a lower bound on C_{max} for the six possible schedules that lie beyond this node, namely 2134, 2143, 2314, 2341, 2413 and 2431. Consideration of this bound will tell us whether to explore this branch of the tree any further. First we have to discuss the bounds; then we can discuss their use. It is convenient not to have double subscripts, so we write

$$a = p_{i1},\ b = p_{i2},\ \text{and}\ c = p_{i3}\ \text{for all jobs}\ J_i$$

We generalize the problem to n jobs while we develop the bounds. When we are at a node $J_{i(1)}, J_{i(2)}, \ldots, J_{i(k)}$. XX . . . XX. K jobs $\{J_{i(1)}, J_{i(2)}, \ldots, J_{i(k)}\}$ have been assigned to the first K positions in the processing sequence. Let A = $(J_{i(1)}, J_{i(2)}, \ldots, J_{i(k)})$ be the subsequence of jobs already assigned and U the set of (n – K) unassigned jobs. For K = 1, 2, . . ., K we let $\alpha_{i(k)}$ be the completion time of job $J_{i(k)}$ on machine 1, $\beta_{i(k)}$ its completion time on machine 2, and $\gamma_{i(k)}$ its completion time on machine 3. $\alpha_{i(k)}$, $\beta_{i(k)}$, and $\gamma_{i(k)}$ are calculated recursively. This is the same as expressing the Gantt chart in equation form by using the two requirements of generating a Gantt chart for a given schedule. First: a job cannot start on a machine until it has finished on the previous machine; second: it cannot start on that machine until the previous job has finished on it.

$$\begin{aligned} &\alpha_{i(1)} = a_{i(k)};\ \alpha_{i(k)} = \alpha_{i(k-1)} + a_{i(k)} \\ &\beta_{i(1)} = a_{i(1)} + b_{i(1)};\ \beta_{i(1)} = \max\{\alpha_{i(k)}, \beta_{i(k-1)}\} + b_{i(k)}; \\ &\gamma_{i(k)} = a_{i(1)} + b_{i(1)} + c_{i(1)};\ \gamma_{i(k)} = \max\{\beta_{i(k)}, \gamma_{i(k-1)}\} + c_{i(k)}; \end{aligned} \qquad \text{Equ. 10.5}$$

These equations define the completion times of the last placed job on all three machines. This general Gantt chart is shown in Figure 10.14.

To develop the lower bound we now consider three possibilities that are the best possible. How quickly can we complete the remaining jobs in U if we compact the processing on each of the three machines in turn? Assume that the processing on machine 1 is continuous. (This is always true for any semi-active schedule). But suppose that the last job in the schedule, $J_{i(n)}$ does not have to wait for either machine 2 or machine 3 to be free before being processed there. Because $C_{max} = C_{i(n)}$, we have the following.

$$C_{max} = \alpha_{i(k)} + \sum_{J_i \text{ in } U} a_{i+}\left(b_{i(n)} + c_{i(n)}\right) \qquad \text{Equ. 10.6}$$

Choosing $J_{i(n)}$ to have the least total processing on machines 2 and 3, all schedules must have

$$C_{max} \geq \alpha_{i(k)} + \sum_{J_i \text{ in } U} a_{i+} min_{J_i \text{ in } U}\left(b_{i(n)} + c_{i(n)}\right) \qquad \text{Equ. 10.7}$$

It is conceivable that processing on machine 2 is continuous. Then there must be no idle time while it waits for jobs still being processed on machine 1. Also

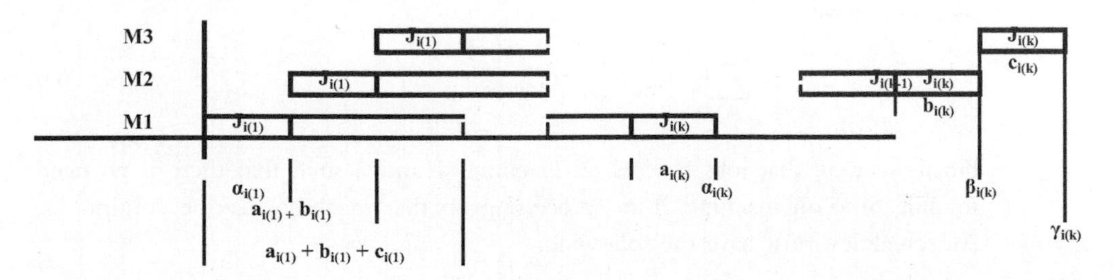

Figure 10.14 General Gantt Chart for Three Machines

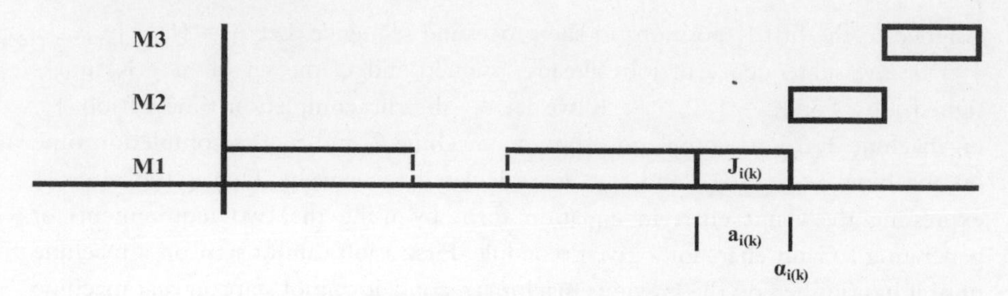

Figure 10.15 Gantt Chart for the First Bound

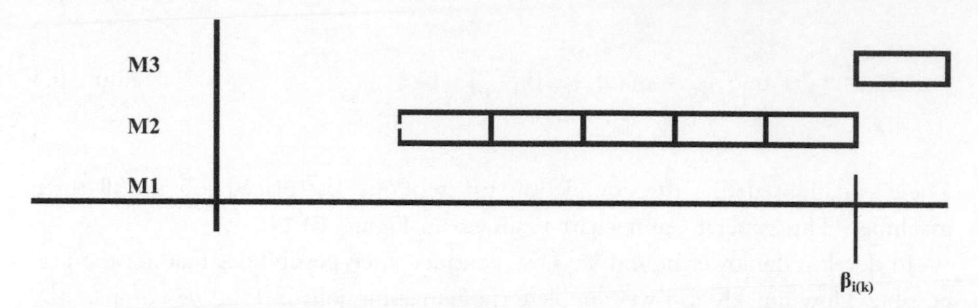

Figure 10.16 Gantt Chart for the Second Bound

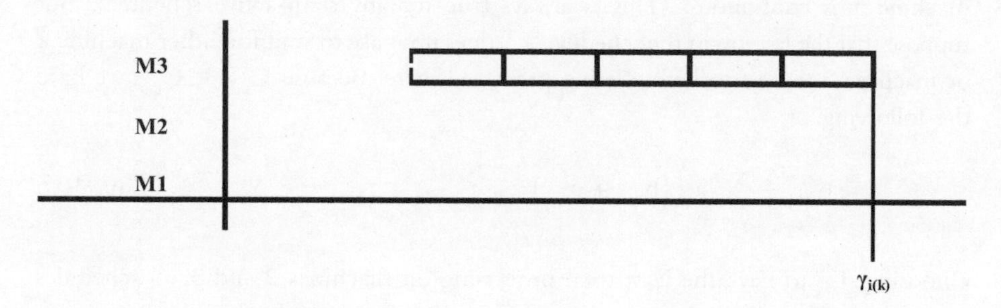

Figure 10.17 Gantt Chart for the Third Bound

suppose that the last job in the schedule, $J_{i(n)}$ can complete on machine 3 without needing to wait for $J_{i(n-1)}$ to complete. We then have the following.

$$C_{max} = \beta_{i(k)} + \sum_{J_i \text{ in } U} b_i + c_{i(n)} \qquad \text{Equ. 10.8}$$

Choosing $J_{i(n)}$ to have the least possible processing time on machine 3, all schedules must have

$$C_{max} \geq \beta_{i(k)} + \sum_{J_i \text{ in } U} b_i + \min_{J_i \text{ in } U} \left(C_{i(n)}\right) \qquad \text{Equ. 10.9}$$

Finally, assume that jobs process on machines 1 and 2 such that there is no need for idle time on machine 3; i.e., processing on that machine may be continuous. All schedules must have the following:

$$C_{max} \geq \gamma_{i(k)} + \sum_{j_i \text{ in } U} C_i \qquad \text{Equ. 10.10}$$

Because no schedule can do better than any of expressions in Equations 10.8, 10.9, and 10.10, no schedule can do better than the maximum of these. The lower bound therefore is:

$$lb(A) = \max\left\{\begin{array}{l}\alpha_{i(k)} + \sum_{J_i \text{ in } U} a_{i+} \min_{J_i \text{ in } U}\left(b_{i(n)} + c_{i(n)}\right),\ \beta_{i(k)} \\ + \sum_{J_i \text{ in } U} b_i + \min_{J_i \text{ in } U}\left(C_{i(n)}\right), \gamma_{i(k)} + \sum_{J_i \text{ in } U} C_i\end{array}\right\} \qquad \text{Equ. 10.11}$$

The lower bound may be strictly less than the make span of all schedules beginning with the subsequence A. In deriving the three components of the lower bound we have assumed that there needn't be waiting or idle time in the schedule. This is rarely so. A lower bound only indicates that no smaller C_{max} is possible. In no way does this say that this is the optimum. Almost always the optimal C_{max} will be greater than the lower bound. As we explore the tree and have more information about the branch as we add jobs, the bounds often get worse, showing that the bound calculated at the initial node was not tight.

The solution of a $4/3/P/C_{max}$ problem follows. The processing times are shown. All the ready times are zero. Note that this problem does not satisfy the Johnson condition for optimality. If we apply it anyway, we get the sequence (2314) with a C_{max} of 31. But here we are interested in getting the optimum. For convenience in calculating $\min_{J(i) \text{ in } U} \{b_i + c_i\}$ and $\min_{J(i) \text{ in } U} \{c_i\}$ for various sets U, $\{b_i + c_i\}$ and c_i are shown in increasing order in Figure 10.18.

We begin by calculating lower bounds on C_{max} at all the nodes on the branch of the elimination tree leading to the processing sequence 1234; i.e., we travel from XXXX to lXXX to 12XX and finally to 1234. Naturally, at the final node 1234 we calculate C_{max} itself and not a lower bound. Thus we obtain:

At 1XXX $A = (J_1), U = \{J_2, J_3, J_4\}$.

$\alpha_1 = 1, \beta_1 = 9, \gamma_1 = 13$.

$lb(A) = \max\{1 + 11 + 9, 9 + 15 + 2, 13 + 15\} = 28$.

Job	Processing Times		
J_i	a_i	b_i	c_i
1	1	8	4
2	2	4	5
3	6	2	8
4	3	9	2

J_i	b_i+c_i		J_i	c_i
2	9		4	2
3	10		1	4
4	11		2	5
1	12		3	8

Figure 10.18 Data for the Branch and Bound Example

At 12XX $A = (J_1, J_2), U = \{J_3, J_4\}$.
$\alpha_2 = 1 + 2 = 3, \beta_2 = \max\{3, 9\} + 4 = 13$,
$\gamma_2 = \max\{13, 13\} + 5 = 18$.
$lb(A) = \max\{3 + 9 + 10, 13 + 11 + 2, 18 + 10\} = 28$.

At 1234 $A = (J_1, J_2, J_3, J_4), U = \{\ \}$.
$\alpha_3 = 3 + 6 = 9, \beta_3 = \max\{9, 13\} + 2 = 15$,
$\gamma_3 = \max\{15, 18\} + 8 = 26$.
$\alpha_4 = 9 + 3 = 12, \beta_4 = \max\{12, 15\} + 9 = 24$,
$\gamma_4 = \max\{24, 26\} + 2 = 28$.

Hence C_{max} for schedule 1234 $= \gamma_4 = 28$.

Note that in this instance the Johnson approximation did not produce a schedule as good as the smallest of these bounds and we have to continue searching. However, we do know that 28 is better than 31 and becomes our new basis for comparison, i.e., our trial schedule. We will further explore the elimination tree comparing the lower bounds at each node with the value of C_{max} for the current trial schedule. If the lower bound at a node is greater than or equal to this we know that we cannot improve upon the trial schedule by exploring that branch further and hence we eliminate that node and all nodes beyond it in the branch. If the lower bound at a node is less than C_{max} of the trial schedule, we cannot eliminate the node and must explore the branch beyond it. If we arrive at a final node and find that the schedule there has C_{max} less than that of the trial, then this schedule becomes the new trial schedule. Eventually we will have eliminated or explored all the nodes and the trial schedule that remains must be optimal.

Back to our problem and applying this process, Figure 10.19 shows the one branch of the elimination tree that we have explored so far. The numbers by the nodes are the lower bounds. Now the lower bound at 12XX is 28, which is no less than C_{max} of the trial. Thus we cannot improve upon the trial 1234 by exploring the branch to 1243. Similarly the lower bound at node 1XXX is also 28 and we may eliminate the branches to nodes 13XX and 14XX and beyond. Thus we have arrived at the stage shown in Figure 10.20.

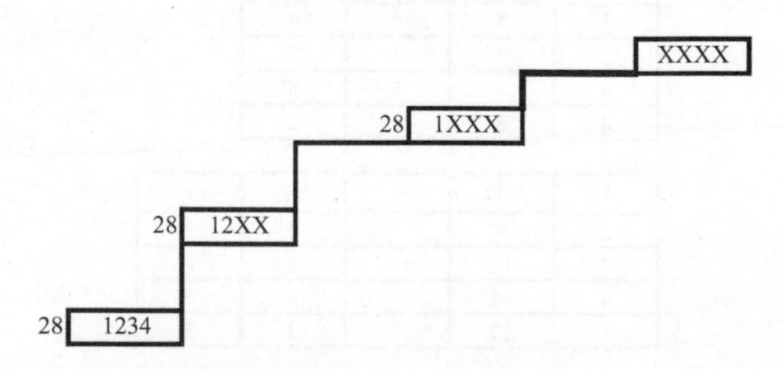

Figure 10.19 The Branch of the Elimination Tree Explored First

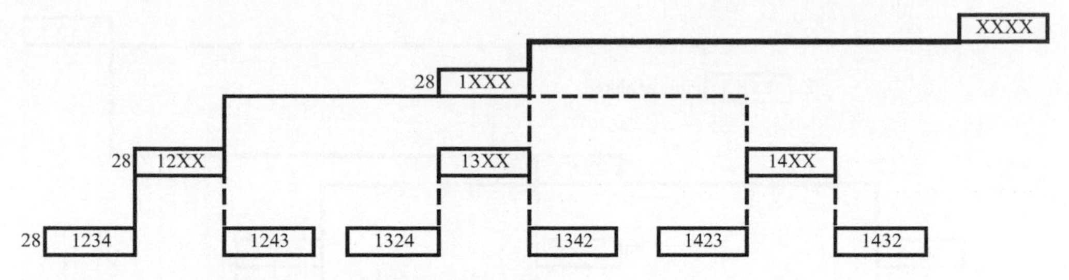

Figure 10.20 All Nodes Beyond 1XXX Fully Explored

Next we explore the branch XXXX to 2XXX and beyond. We make the following calculations:

At 2XXX $A = (J_2), U = \{J_1, J_3, J_4\}$.
$\alpha_1 = 2, \beta_1 = 6, \gamma_1 = 11$.
$lb(A) = \max\{2 + 10 + 10, 6 + 19 + 2, 11 + 14\} = 27$.

This lower bound is less than 28, the C_{max} of the trial. Therefore we must explore the branches beyond the node 2XXX.

At 21XX $A = (J_2), U = \{J_1, J_3, J_4\}$.
$\alpha_1 = 2 + 1 = 3, \beta_1 = \max\{3, 6\} + 8 = 14, \gamma_1 = \max\{14, 11\} + 4 = 18$.
$lb(A) = \max\{3 + 9 + 10, 14 + 11 + 2, 18 + 10\} = 28$.

At 23XX $A = (J_2, J_3), U = \{J_1, J_4\}$.
$\alpha_3 = 2 + 6 = 8, \beta_3 = \max\{8, 6\} + 2 = 10, \gamma_3 = \max\{10, 11\} + 8 = 19$.
$lb(A) = \max\{8 + 4 + 11, 10 + 17 + 2, 19 + 6\} = 29$.

At 24XX $A = (J_2, J_4), U = \{J_1, J_3\}$.
$\alpha_4 = 2 + 3 = 5, \beta_4 = \max\{5, 6\} + 9 = 15, \gamma_4 = \max\{15, 11\} + 2 = 17$.
$lb(A) = \max\{5 + 7 + 10, 15 + 10 + 4, 17 + 12\} = 29$.

The lower bounds at each of these nodes are not less than the value of C_{max} for the trial schedule; so there is no need to explore these branches further. Thus we arrive at the situation shown in Figure 10.21.

Next we explore the branches beyond 3XXX.

At 3XXX $A = (J_3), U = \{J_1, J_2, J_4\}$.
$\alpha_3 = 6, \beta_3 = 8, \gamma_3 = 16$.
$lb(A) = \max\{6 + 6 + 9, 8 + 21 + 2, 16 + 11\} = 31$.

Because this lower bound is greater than 28, we explore no further along this branch. Finally we look at the branches beyond 4XXX.

At 4XXX $A = (J4), U = \{J1, J2, J3\}$.
$\alpha 4 = 3, \beta 4 = 12, \gamma 4 = 14$.
$lb(A) = \max\{3 + 9 + 9, 12 + 14 + 4, 14 + 17\} = 31$.

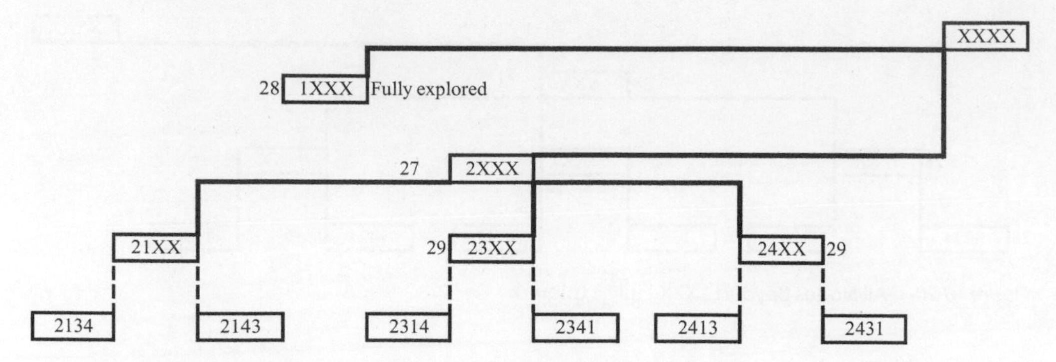

Figure 10.21 All Nodes Beyond 1XXX and 2XXX Fully Explored

Because this lower bound is greater than 28, there is no need to consider the branches beyond 4XXX.

We have now explored the entire elimination tree and shown that no schedule can have a C_{max} less than 28. The current trial schedule 1234 is optimal. The complete elimination tree for the above is given in Figure 10.22. In our solution here it happened that the first schedule we evaluated remained the trial for the entire solution. This happened purely by chance. Had we explored the branch XXXX to 4321 first, the trial schedule would have changed in the course of the solution. Try it.

The search procedure we used is called a depth-first search. In our search of the tree we selected a branch and systematically worked down it until we had either eliminated it on the grounds of a lower bound or had reached its final node, which had either become the trial schedule or was eliminated. This search strategy has the advantage that the computer (or we) need only remember $\alpha_{i(k)}$, $\beta_{i(k)}$, $\gamma_{i(k)}$ and lb(A) for the nodes in the branch currently being searched. Because there are at most (n − 1) such nodes in an n-job problem, this search procedure requires little memory. Note that at no time was it necessary to remember the value of $\alpha_{i(k)}$, $\beta_{i(k)}$, $\gamma_{i(k)}$ and lb(A) at more than 3 = 4 − 1 nodes. Because of the systematic way in which we searched the tree, once the immediate need for these quantities at a node was past we knew that we would not need them again.

However it might need a lot of computation. We were lucky. We might have needed to explore the branches much further before they were eliminated. An alternative strategy is a frontier search or branch-from lowest-bound. As the latter name suggests, in this we always branch from a node with the current lowest lower bound. It is easiest to follow this procedure through an example. We will solve the $4/3/F/C_{max}$ problem again, this time by a frontier search. You can verify our calculations. First we branch to each of the four nodes 1XXX, 2XXX, 3XXX, 4XXX simultaneously and calculate the lower bounds. We obtain the result shown in Figure 10.23. The node 2XXX has the lowest lower bound so we branch from this to nodes 21XX, 23XX, and 24XX to obtain the result shown in Figure 10.24. Here we find two nodes 1XXX and 21XX share the same lower bound. Picking

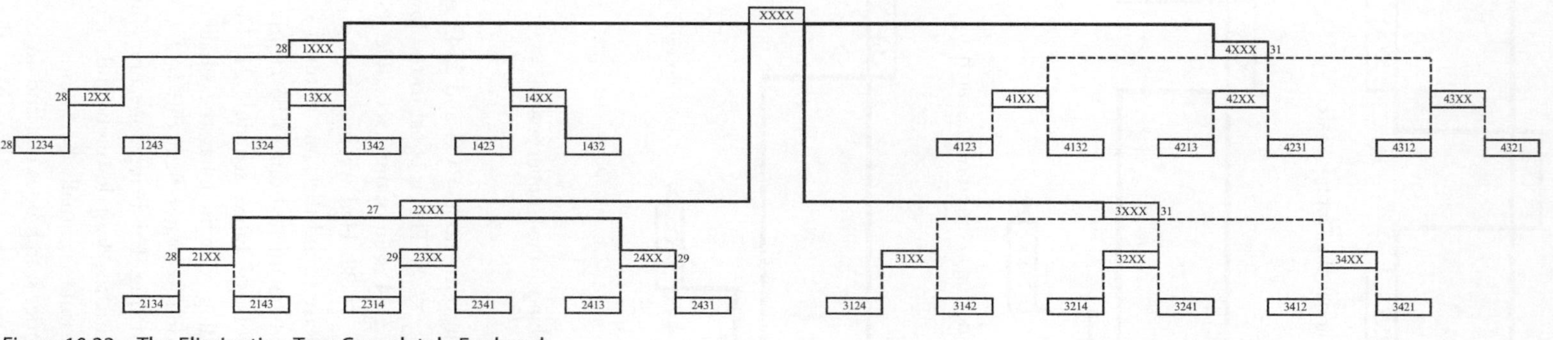

Figure 10.22 The Elimination Tree Completely Explored

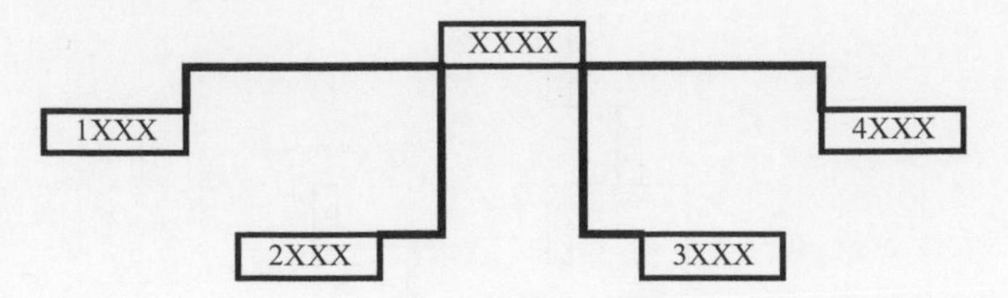

Figure 10.23 The First Branching in the Frontier Search

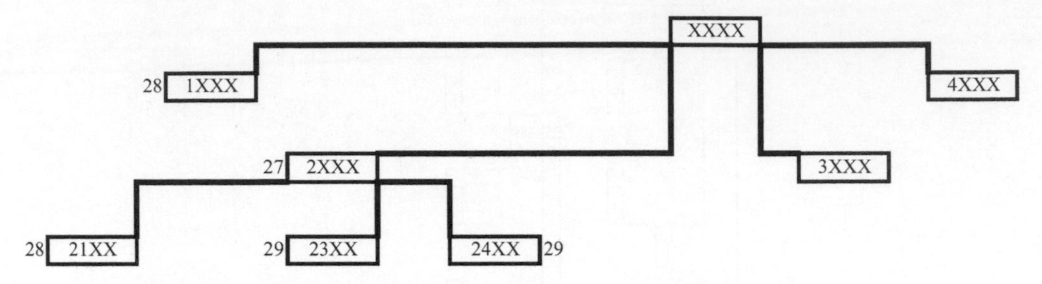

Figure 10.24 The Second Branching in the Frontier Search

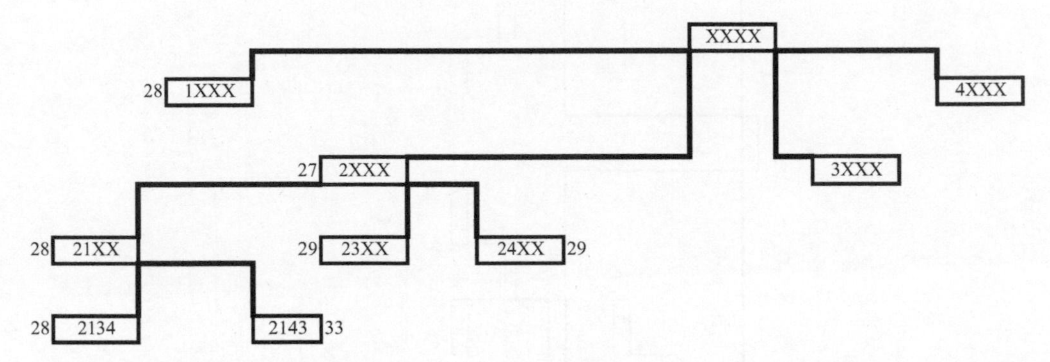

Figure 10.25 The Third Branching in the Frontier Search

21XX arbitrarily we branch to the nodes 2134 and 2143 as shown in Figure 10.25. The schedule 2134 has C_{max} = 28. This is equal to the lowest of the lower bounds obtained in the tree. Thus 2134 is an optimal schedule. (Schedules 2134 and 1234 are both optimal with C_{max} = 28. Had we branched at 1XXX instead of 21XX the frontier search would have found the same optimal schedule as the depth-first search.) Had the node 2134 not turned out to be clearly an optimal schedule, we would have proceeded as follows. Either schedule 2134 or 2143 would have been selected as the trial according to which has the smaller completion time. Then those nodes with lower bounds greater than this best yet C_{max} would have been eliminated from further searching. This being done, branching would have continued from a remaining node with the least lower bound. Continuing in the obvious way, an optimal schedule would eventually be found.

In this example, the frontier search does not find an optimal schedule faster than the depth first search. If you were asked to solve the above problem with a depth-first search that sweeps across the tree from right to left rather than from left to right. You would find that in doing so you need more calculation than in the frontier search. In general, a frontier search will require less calculation than a depth-first search. The frontier search chooses which branch to explore next in a more intelligent fashion and so usually finds an optimal solution faster.

Before closing this section, notice that the above example does not conform to the conditions under which Johnson's algorithm may be applied. If we apply the algorithm nonetheless, we obtain the schedule 2314, which by the lower bound at 23XX must have $C_{max} \geq 29$ and so cannot be optimal. Therefore it is clear that the conditions are not redundant, but are necessary to the success of that algorithm. But you should also notice that Johnson's algorithm generated a schedule very close to the optimum. It generally will as long as the interior machine's processing times are not too large.

Some General Points About the Branch and Bound Approach

The number of operations required, and hence the time required, to solve a problem by branch and bound is unpredictable, whatever search strategy is used. It might happen that the procedure has to explore fully virtually every node, in which case it would take as long as complete enumeration. It might take longer because branch and bound involves more computation per node than complete enumeration. In general branch and bound does perform a great deal better than complete enumeration. But it should not be assumed from this that it can solve any problem in practice. Theoretically, like dynamic programming, it always finds an optimal solution, but it may take prohibitively long to do so.

The search strategy used is an important determinant of the time required to solve a problem. A frontier search generally finds an optimal schedule faster than a depth-first search. The quality of the lower bounds is just as important. If the lower bound at a node is good, i.e., not much less than the least value of the performance measure there, then the procedure generally finds an optimal schedule after examining fewer nodes than it would with poorer bounds. Good bounds eliminate nodes high in the tree, thus reducing the search substantially. Consider the example above. At each node zero is a lower bound, albeit a much poorer one than Equation 10.5. When zero is used, every branch must be explored. Clearly the better bound is to be preferred. Despite these comments, it must not be assumed that the better the lower bounds the faster will branch and bound find the optimal schedule. It depends upon how long the calculation takes on each node. It may be faster overall to calculate poor lower bounds quickly and examine many nodes in the search than to calculate good lower bounds slowly and examine far fewer nodes. Empirical investigations have shown that the extra effort required to calculate good lower bounds is time well spent, the rate of eliminating nodes increases more than enough to compensate. Rinnooy Kan, Lageweg, and Lenstra (1975) have suggested using different bounds in different parts of the elimination tree. When a node high in the tree is eliminated, many subsequent nodes are eliminated at the same time. Lower down in the tree each node has fewer subsequent nodes to be eliminated along with it. It may be best to spend time calculating very good bounds in the upper levels of the tree, and calculate quick but poor bounds in the lower levels.

Besides the choice of search strategy and lower bounds, there are other ways in which we can try to increase the speed of finding an optimal solution. First, we can prime the procedure with a near-optimal schedule as the first trial. The better the first trial the more nodes we may expect to be eliminated in the early stages. Assuming that we can find a near optimal schedule, we may save ourselves a lot of computation. That assumption, of course, begs the question: how do we find a near optimal schedule? The answer is to use heuristic methods, which are covered in later chapters. These are part of a family of sensible 'rules of thumb' that we expect from experience and intuition to produce good, if not optimal, schedules. For example, it is known that, if Johnson's algorithm is applied to an $n/3/F/F_{max}$ problem for which the conditions do not hold, then a good schedule is obtained. We might use this to generate a first trial schedule for a branch and bound solution.

Second, we may consider employing dominance conditions. We have already suggested their use in dynamic programming; their use in branch and bound is much better investigated. Suppose we have a set of conditions such that, when they apply to a pair of nodes, we may deduce that all the schedules at one node can do no better than the best schedule at the other. Then clearly we may eliminate the first node from further consideration. For example, consider an $n/3/F/F_{max}$ problem and refer to the notation used in developing the recursions in equation 10–4. Suppose that we compare two nodes $J_{i(1)}, J_{i(2)}, \ldots, J_{i(k)}, XX \ldots X$ and $J_{j(1)}, J_{j(2)}, \ldots, J_{j(k)}, XX \ldots X$ at which the same K jobs have been assigned to the first K positions, but in different orders; i.e., the nodes share the same set of unassigned jobs. Suppose further that the following conditions hold:

$$\alpha_{i(k)} \geq \alpha_{j(k)}, \beta_{i(k)} \geq, \beta_{j(k)} \text{ and } \gamma_{i(k)} \geq \gamma_{j(k)}$$

That is, the completion time of the Kth job on each machine is no earlier under the first subsequence than it is under the second. Then, however we complete the processing of the unassigned jobs at the first node, we shall do at least as well with the same completion at the second. Thus we may eliminate the first node from the search. Checking for such dominance conditions during the search of a tree may take considerable computation and storage. Nonetheless, because they may eliminate many nodes before lower bounding arguments can do so, their use may curtail the search sufficiently that, overall, a reduction in computational requirements is obtained. This has been found in practice. Baker (1975), Rinnooy Kan, Lageweg, and Lenstra (1975), and Lageweg, Lenstra, and Rinnooy Kan (1978) all report that careful inclusion of dominance conditions in branch and bound can lead to improvements in performance.

A final method for saving computation is to accept a sub-optimal solution. Suppose we agree that any solution within 10% of the optimal would be satisfactory; i.e., if Y_{opt} is an optimal schedule and y is a schedule such that $c(y) \leq 1.10\, c(y_{opt})$, then we would accept Y as a solution. Now suppose that at some stage of a branch and bound search we have found a trial schedule $\gamma^\star$. Then we may eliminate all nodes with lower bounds greater than $c(Y^\star)/1.10$. For, if one of those nodes were to lead to an optimal schedule, then it would have to be within 10% of our current trial. Thus when looking for a sub-optimal solution we may eliminate nodes faster than if we were only prepared to accept an optimal schedule. Kohler and Steiglitz (1976) report that the savings in computation that result may be dramatic.

11 OPTIMIZATION

Introduction

A considerable amount of literature suggests solving scheduling problems by stating them as integer programs. These recast problems may be solved by standard algorithms, which have been developed for solving general mathematical problems. Translating back, we obtain optimal schedules. This sounds like a very promising approach, but in practice it has not proven to be. The standard mathematical programming algorithms are practically applicable to quite large problems; the recast practical scheduling problems can be substantially larger. Thus the recasting is just that: the inherent difficulties are rephrased, but not into a more tractable form. Empirically this confirms that scheduling problems are in general very difficult and do not just appear to be so.

We will translate an $n/m/P/C_{max}$ scheduling problem into a mixed integer program. Once translated, commercially available software can solve it. The body of theory and algorithms called mathematical programming deals with problems of the following kind:

> Minimize $f(x_1, x_2, \ldots, x_l)$ with respect it $x_1, x_2, \ldots, x_l$ subject to the constraints:
>
> $g_1(x_1, x_2, \ldots, x_m) \leq b_1$
> $g_2(x_1, x_2, \ldots, x_m) \leq b_2$
> ...
> ...
> $g_k(x_1, x_2, \ldots, x_m) \leq b_k$

Mathematical programming is a family of techniques for optimizing a function subject to constraints on the independent variables. In scheduling we wish to optimize a performance measure subject to technological constraints on the allowable processing order. Thus it is not surprising that, given some effort, the latter can be translated into the former type of problem; both concern optimization under constraints.

In Integer programming or, to be strictly correct, mixed integer programming, some of the independent variables are constrained to be integral. Frequently they are constrained to 0 or 1 (binary) and are used to indicate the absence or presence of some property. The functions f, $g_1, g_2, \ldots, g_k$ are linear. (Non-linear problems may be solved with considerably more difficulty). Thus the standard problem takes the form:

$$\begin{aligned}
&\text{Minimize } c_1x_1 + c_2x_2 + c_3x_3 + \ldots c_lx_l \text{ subject to}\\
&\quad g_{11}x_1 + g_{12}x_2 + \ldots g_{1m}x_m \leq b_1\\
&\quad g_{21}x_1 + g_{22}x_2 + \ldots g_{2m}x_m \leq b_2\\
&\quad \ldots\\
&\quad \ldots\\
&\quad g_{k1}x_1 + g_{k2}x_2 + \ldots g_{km}x_m \leq b_k
\end{aligned}$$

and some of the $x_1, x_2, \ldots, x_m$ are limited to integral values. Some of the inequality constraints may be replaced by strict equality. Methods of solving such problems are reviewed in most introductory texts on operation research, e.g., Hillier and Lieberman (1990). By and large, those methods can be classified as either implicit enumeration, in particular branch and bound, or cutting plane. Both require much computation. Both are based on the properties of integer programs in general and pay no regard to the particular properties of the problem being solved. As a result they tend to take longer to find a solution than implicit enumeration algorithms designed specifically for a particular class of problems. For example, consider two methods of solving a scheduling problem. First, we may translate it into an integer program and solve that by branch and bound with the bounds based upon general integer programming theory. Or, second, we may tackle the problem directly by branch and bound with bounds based upon our knowledge of the physical properties of schedules. The lower bounds found in the first case are usually poorer than those found in the second case and the branch and bound search correspondingly longer. So, as we have said, it is better to approach scheduling problems directly rather than indirectly via integer programming.

Wagner's Integer Programming Form of $n/m/P/C_{max}$ Problem

Wagner (1959) introduced the following integer programming formulation of the permutation flow shop problem $n/m/P/C_{max}$. We assume that all ready times are zero. The subscript i will refer to the job J_i the subscript j to the machine M_j; and the subscript k to the kth position in the processing sequence. Thus the jobs will be processed through each machine in the order $(J_{i(1)}, J_{i(2)}, \ldots, J_{i(n)})$. Moreover, because the permutation flow shop is a subset of the flow shop, each job has the same technological constraints $(M_1, M_2, \ldots, M_m)$. To model this problem we introduce n^2 variables constrained to take the values 0 or 1:

$$X_{ik}= \begin{Bmatrix} 1, \text{ if } J_i \textit{ is scheduled in the kth position of the processing sequence;} \\ 0, \textit{ otherwise} \end{Bmatrix}$$

The constraints that must be obeyed by these variables are:

$$\sum_{i=1}^{n} X_{ik} = 1 \text{ for } k = 1, 2, \ldots, n \qquad \text{Equ. 11.1}$$

That is, exactly one job is scheduled in the kth position and

$$\sum_{k=1}^{n} X_{ik} = 1 \text{ for } i = 1, 2, \ldots, n \qquad \text{Equ. 11.2}$$

That is, each job is scheduled in exactly one position. These constraints force each X_{ik} to take the values 0 and 1 only, provided we restrict the X_{ik} to non-negative integers. Excel's Solver allows the definition of a binary variable that satisfies both conditions simultaneously.

Next we introduce two sets of non-negative real variables: I_{jk} and W_{jk}. I_{jk} is the idle time of machine M_j between the completion of $J_{i(k)}$ in the processing sequence and the start of $J_{i(k+1)}$. Thus I_{jk} is defined for j = 1, 2, . . ., m and k = 1, 2, . . ., n – 1. Because there is no idle time on the first machine, $I_{1k} = 0$ for k = 1, 2, . . ., n – 1. See Figure 11.1.

The waiting time W_{jk} is the time that $J_{i(k)}$ must spend between completion on M_j and starting to be processed on M_{j+1}. Thus W_{jk} is defined for j = 1, 2, . . ., m – 1 and k = 1, 2, . . ., n. Because there is nothing to delay the first job processed, $W_{j1} = 0$ for j = 1, 2, . . ., m – 1. Again see Figure 11.1.

The times t_{jk} between completion of the job $J_{i(k)}$ on M_j and the start of the job $J_{i(k+1)}$ on M_{j+1} must be well defined. From Figure 11.2 we see that

$$t_{jk} = I_{jk} + p_{i(k+1)j} + W_{j,k+1} = W_{jk} + p_{i(k)j+1} + I_{j+1,k} \qquad \text{Equ. 11.3}$$

We must express $p_{j(k+1)j}$ and $p_{j(k+1)j}$ in terms of the X_{ik}, because we do not know explicitly how to find the subscripts i(k + 1) and i(k). Now, because X_{jk+1} is zero except for i = i(k + 1) when it is 1, it follows that

$$p_{i(k+1)j} = \sum_{i=1}^{n} X_{i,k+1} p_{ij} \qquad \text{Equ. 11.4}$$

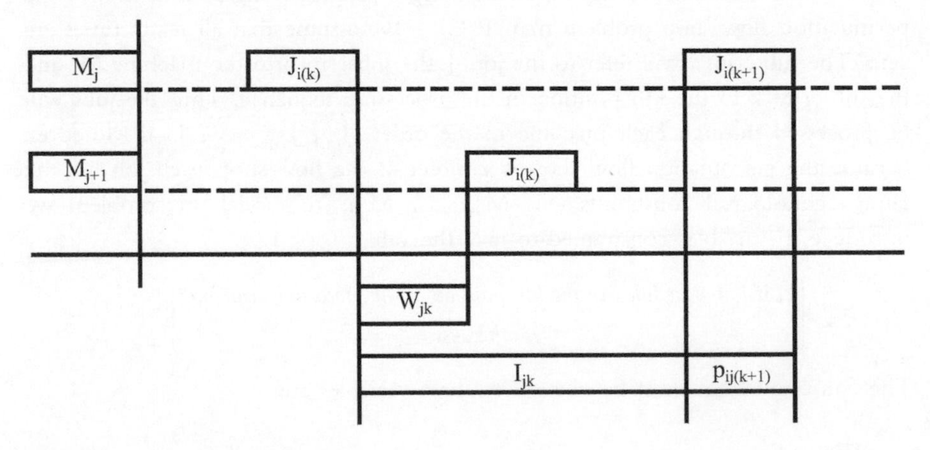

Figure 11.1 Definition of I_{jk} and W_{jk}

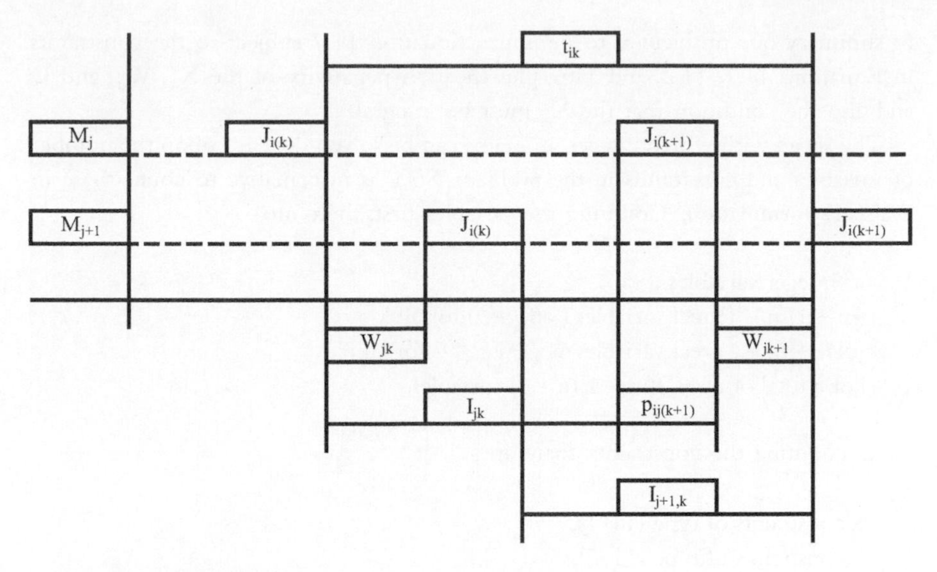

Figure 11.2 Definition of t_{jk}

Similarly,

$$p_{i(k)j+1} = \sum_{i=1}^{n} X_{i,k} p_{ij+1} \qquad \text{Equ. 11.5}$$

We rewrite Equation 11.3 as:

$$I_{jk} + \sum_{i=1}^{n} X_{i,k+1} p_{ij} + W_{j,\,k+1} - W_{jk} + \sum_{i=1}^{n} X_{i,k} p_{ij+1} - I_{j+1,k} = 0 \qquad \text{Equ. 11.6}$$

These constraints hold for j = 1, 2, . . ., m − 1, k = 1, 2, . . ., (n − 1). The constraints of Equation 11.6 not only ensure that the times t_{jk} are well defined, but also that the technological constraints of the flow shop are obeyed.

The constraints in Equations 11.1, 11.2, and 11.6, together with the demands that X_{ik} are non-negative integers and that W_{jk} and I_{jk} are non-negative real numbers, form the entire constraint set. The objective function that we wish to minimize is defined as follows.

Minimizing C_{max} is equivalent to minimizing the idle time on the last machine. The total idle time on M_m is given by the sum of the inter-job idle times I_{mk} plus the idle time that must occur before job J_{i1} can start processing on M_m. Thus we seek to minimize $\sum_{k=1}^{n-1} I_{mk} + \sum_{j-1}^{m-1} p_{i(1)j}$. However, we must express the sum $\sum_{j=1}^{m-1} p_{i(1)j}$ in terms of the X_{ik} variables. So we seek to minimize:

$$\sum_{k=1}^{n-1} I_{mk} + \sum_{j=1}^{m-1}\left(\sum_{i=1}^{n} X_{i1} p_{ij}\right) \qquad \text{Equ. 11.7}$$

In summary our problem is to minimize Equation 11.7 subject to the constraints in Equations 11.1, 11.2, and 11.6, plus the non-negativity of the X_{ik}, W_{jk}, and I_{jk} and also the condition that the X_{ik} must be integral.

The speed with which integer programs can be solved depends upon the number of variables and constraints in the problem. So it is informative to count these in Wagner's formulation. Counting the variables first, there are

n^2 integer variables
$(m - 1)(n - 1)$ real variables I_{jk} ($I_{1k} = 0$ for all k)
$(m - 1)(n - 1)$ real variables W_{jk} ($W_{j1} = 0$ for all j)
For a total of $n^2 + 2(m - 1)(n - 1)$ variables.

Next, counting the constraints, there are

n constraints of type (11–1),
n constraints of type (11–2),
$(m - 1)(n - 1)$ constraints of type (11–6),
For a total of $mn + n - m + 1$ constraints.

There are, of course, $n^2 + 2(m - 1)(n - 1)$ non-negativity constraints upon the variables, but these need not be counted explicitly because integer programming algorithms invariably include them implicitly.

Another Formulation

The Wagner model does solve for the correct sequence, but leaves the values of I and W as not unique, and hence not useful for the resulting Gantt chart. Formulating the model using the starting times and the previously derived formulas in Equations 11.4 and 11.5, we can obtain a unique solution.

Define $S_{i(k)j}$ as the starting time of $J_{i(k)}$ on machine j and the corresponding completion time as $C_{i(k)j}$. The two are then related by the processing time:

$$C_{i(k)j} = S_{i(k)j} + p_{i(k)j} = S_{i(k)j} + \sum_{i=1}^{n} X_{i,k} p_{ij} \qquad \text{Equ. 11.8}$$

The relationship between the starting and completion times is given by two conditions (recall the generation of a Gantt chart and the derivation of the bounds for C_{max} in Chapter 10).

The start of a job on a machine must be greater than or equal to the completion time of that job on the previous machine:

$$S_{i(k)j} \geq C_{i(k)j-1} \qquad \text{Equ. 11.9}$$

The start of a job on a machine must be greater than or equal to the completion time of the previous job on the same machine:

$$S_{i(k)j} \geq C_{i(k-1)j} \qquad \text{Equ. 11.10}$$

Our objective then is to minimize the maximum completion time $C_{i(n)m}$ with variables X_{ij} and $S_{i(k)j}$ (Equation 11.7).

12 HEURISTIC APPROACHES

In this chapter we will look at methods that cannot guarantee optimal solutions, but nevertheless have a strong tendency to produce good solutions. These are called heuristics. As such, most of these are applicable to any kind of problem, as long as the problem is formulated for a specific heuristic. They are not only useful in scheduling, but can be used for any kind of combinatorial optimization. The ones that we will look at are not an exhaustive list, but do represent a large percentage of available heuristics. We will cover random pairwise exchanges (PE), genetic algorithms (GA), simulated annealing (SA), the shifting bottleneck heuristic (SB), and Monte Carlo methods (MC).

Random Pairwise Exchange

The idea behind this method is to sequentially modify a sequence of digits by exchanging pairs of positions in the sequence. These exchanges are completely random, so there is no method involved to improve subsequent solutions. Because we are exchanging positions in a sequence, the method is very well suited to problems that can be expressed as such. Good examples of these are single-machine problems and any permutation problems.

It is most instructive to apply any heuristic for which we can obtain an optimal solution through other means. This way we can judge the effectiveness of the heuristic by applying it to a set of sample problems.

Let us start with a one-machine problem with a performance measure of average tardiness ($\overline{\text{F}}$). Sometimes these are referred to as total tardiness or $\sum F$. These are equivalent as the only difference is the division by the number of jobs. If we select small enough examples, such as eight jobs, we can efficiently find optimal solutions with complete enumeration. Figure 12.1 shows the data for one such example.

The optimal solution is 0.25 (as obtained by complete enumeration), while achieving the same result required an average of 20,604 random exchanges, with a maximum of 40,662 in 10 trials. Not a bad performance for a procedure with no inherent intelligence. Note that complete enumeration would not only involve 40,320 measure evaluations as in the random process, but also the generation of the same number of unique sequences. Figure 12.2 displays the rate of convergence of these 10 trials with the dashed line showing an upper bound. This is not meant as proof of the value of random exchanges, but does indicate success for this kind of problem.

Job	1	2	3	4	5	6	7	8
p	2	2	2	3	4	4	5	4
d	11	20	8	6	23	17	10	25
C	2	4	6	9	13	17	22	26
T	0	0	0	3	0	0	12	1

Figure 12.1 One Machine, Minimize Average Tardiness

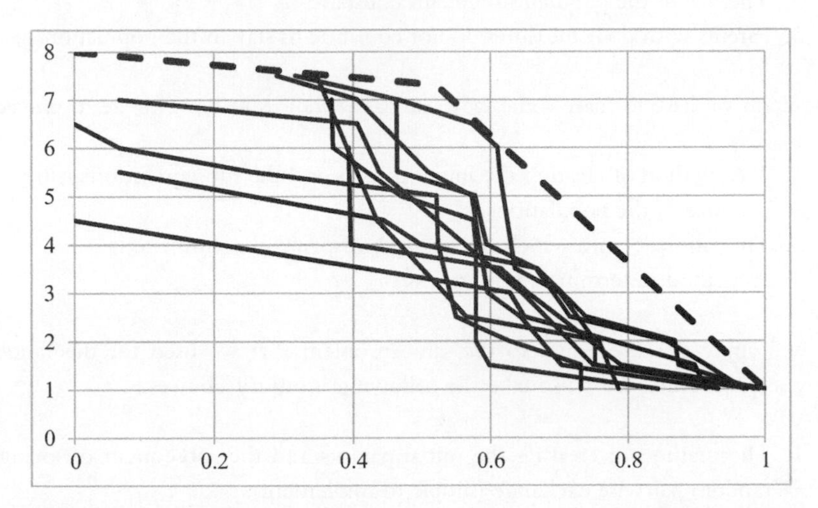

Figure 12.2 Convergence of Random Pairwise Exchange, Ratio of Measure to Optimum vs. Log(Iterations)/Log(All Possible Sequences)

Original	1	2	3	4	5	6	7	8	2
First Exchange	1	2	7	4	5	6	3	8	3
Second Exchange	1	2	7	4	6	5	3	8	2.625
Third Exchange	3	2	7	4	6	5	1	8	2.25

Figure 12.3 First Three Random Exchanges

Figure 12.3 shows a few random exchanges that were performed manually to initiate the process. Note that the improvement or deterioration is unpredictable. Each of these sequences is considered a neighborhood of the previous sequence as defined by random pairwise exchange. Our choice of just exchanging a pair was purely arbitrary; we could have just as easily chosen to exchange pairs of jobs and defined the neighborhood as that. Also, because the exchanges are random, sequences will often be repeated. We will discuss how to deal with this phenomenon after we have looked at the various heuristics, as this is common to many of them.

Genetic Algorithm

Our next heuristic is the genetic algorithm. It is unfortunate that its traditional name is an algorithm, because that tends to imply that it guarantees optima, which of course it does not. It is loosely based on the science of genetics, with some substantial simplifications. These are as follows:

1. A single parent produces an offspring;
2. The size of the population remains constant;
3. Parents with poor measures do not continue to stay in the population.

In addition we have to make some rather arbitrary choices in how we are to proceed:

1. The method of creating the initial parents and the subsequent offspring;
2. The size of the population;
3. The rationale with which parents and offspring are eliminated;
4. A method to terminate the process.

We will apply this procedure to the same problem that we used for the random pairwise procedure. We will make the following arbitrary choices:

1. The method of creating the initial parents and the subsequent offspring—random pairwise exchange (simple to implement);
2. The size of the population—five (large enough to have an effect, small enough to display);
3. The rationale with which parents and offspring are eliminated—replace the worst parent with the best offspring only if it is better;
4. Method of termination—if we did not know the solution we could use a time limit, for example 20 seconds for a problem this size. We will terminate when the optimum is reached and examine the rate of convergence as we did previously.

Figure 12.4 shows the progression of the algorithm for two generations that was created manually, the replacing offspring and the replaced parent are shaded in gray. Figure 12.5 displays the convergence of 10 trials.

One way to compare this convergence to that of purely random exchanges is to note that the most generations it required to reach the optimum was 70 and that there are 10 sequences in each generation, or a total of 700 sequences. The log ratio of this to complete enumeration is 0.61, which is substantially less than the one obtained by random exchange. Clearly, a genetic algorithm is a vast improvement over both random pairwise exchange and complete enumeration.

As we generate sequences it is unavoidable to repeat previously evaluated sequences. Because fewer calculations are required to evaluate a sequence than are required to compare the sequence to all previous sequences, we simply calculate the measure again and ignore the sequence. Some authors have suggested keeping

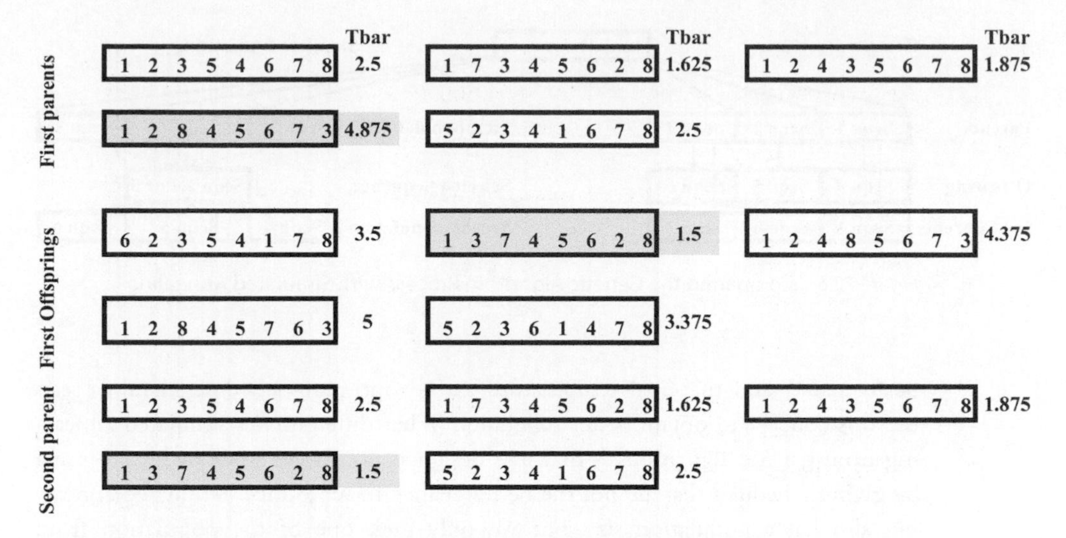

Figure 12.4 Genetic Algorithm from First Generation to Second Generation

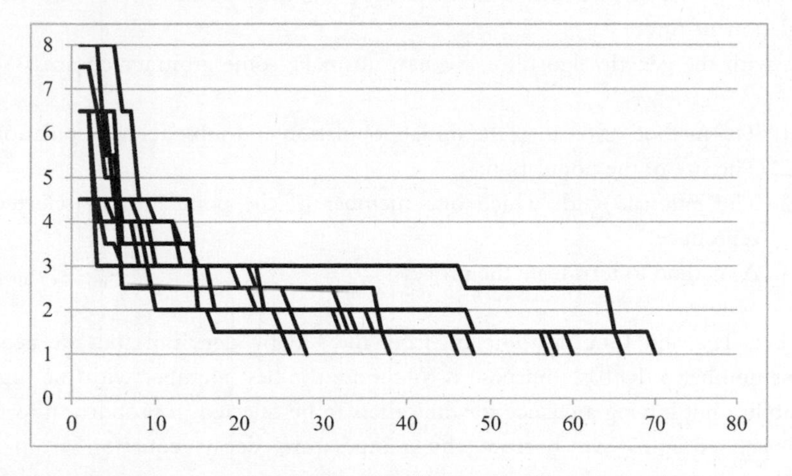

Figure 12.5 Convergence of the Genetic Algorithm, Ratio of Measure to Optimum vs. Number of Generations

a short list of recently visited sequences but this would only be beneficial if substantial calculations are involved in evaluating a measure. This method is referred to as a taboo search.

Simulated Annealing

Our next heuristic that generates many sequences is simulated annealing. It derives its name from the process of cooling molten metal in order to harden it. During this process the temperature is reduced in the material in a decaying manner that eventually results in an equilibrium temperature throughout the body of the material. If we think of the hot initial temperatures as schedules with a poor

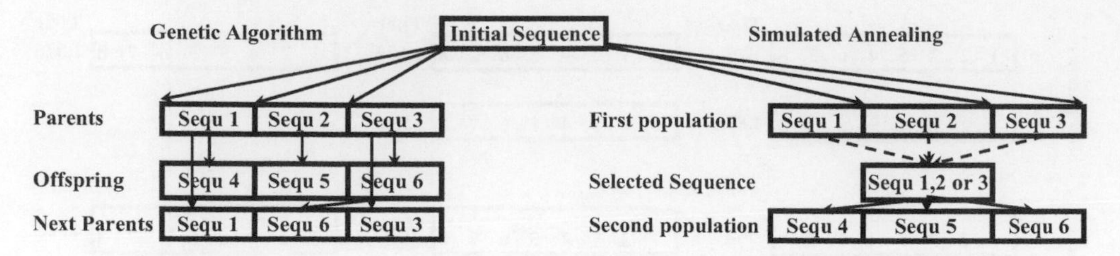

Figure 12.6 Comparing the Genetic Algorithm Process with Simulated Annealing

performance and the final temperature as the optimum schedule, then we can use this concept to obtain better schedules. What differentiates simulated annealing from genetic algorithms is that it is designed not to get stuck at local optima by giving schedules that are not the best a chance to reproduce. Simulated annealing also has a population size, but we only pick one of the population from which to derive the next population. The essence of the process is the method by which we select that one. This difference is illustrated in Figure 12.6 with a population of three.

As with the genetic algorithm, we have to make some arbitrary choices:

1. The method of creating the initial population and subsequent populations;
2. The size of the population;
3. The rationale with which one member of the population is selected to reproduce;
4. A method to terminate the process.

Numbers 1, 2, and 4 are not different from the genetic algorithm, but we need to discuss number 3. Its basic purpose is to choose the best member with the highest probability, but leaving a chance for the others to be selected in proportion to their goodness. For this we borrow the temperature decay equation from real annealing:

$$p_j = e^{-\left(D_j - D_{Best}\right)^{k}/K} \qquad \text{Equ. 12.1}$$

Where D is the measure, p is the probability, and k/K is a convergence constant. The smaller the k/K ratio, the more spread there will be in the probabilities and more potential diversification in the search. Figure 12.7 shows three tables where the probabilities are calculated from our example for two different k/K ratios. The measures here are the sum of tardiness, rather than the average, because they are equivalent.

The method was applied to the previous problem with k/K = 0.1. The convergence for 10 trials is shown in Figure 12.8. In this case it is somewhat faster than the genetic algorithm. However, most of the literature on the subject is indicative of the fact that this is very much dependent on the particular problem and is no proof that simulated annealing is better in all cases.

k/K=	0.1				
Best=	6	D_j-D_{Best}	$e^{-(D_j - D_{Best})k/K}$	p_j Normalized	Cumulative
Schedule 1	12	6	0.549	0.190	0.190
Schedule 3	6	0	1.000	0.346	0.536
Schedule 5	14	8	0.449	0.155	0.691
Schedule 4	21	15	0.223	0.077	0.768
Schedule 2	10	4	0.670	0.232	1.000
			2.892	1	

k/K=	0.9				
Best=	6	D_j-D_{Best}	$e^{-(D_j - D_{Best})k/K}$	p_j Normalized	Cumulative
Schedule 1	12	6	0.005	0.004	0.004
Schedule 3	6	0	1.000	0.968	0.973
Schedule 5	14	8	0.001	0.001	0.974
Schedule 4	21	15	0.000	0.000	0.974
Schedule 2	10	4	0.027	0.026	1.000
			1.033	1	

Figure 12.7 Calculating Probabilities for Simulated Annealing

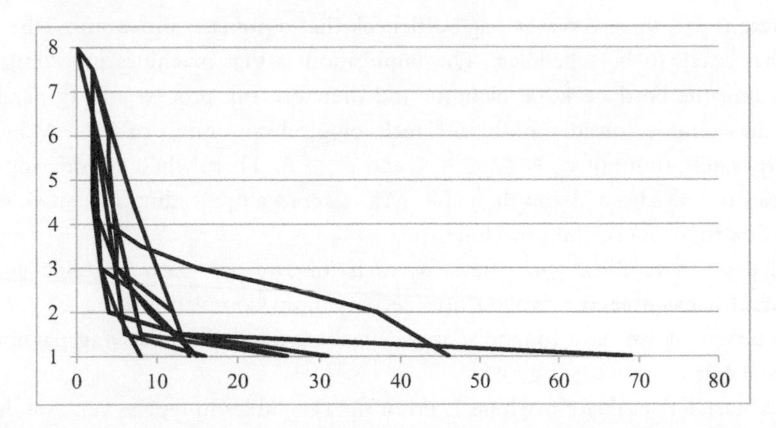

Figure 12.8 Convergence of Simulated Annealing, Ratio of Measure to Optimum vs. Number of Generations

Theory of Constraints

This approach was popularized in the 1980s by E.M. Goldratt (Goldratt, 1990). This theory states that in all likelihood a system is dominated by one bottleneck, that is, one machine that is the busiest or most heavily loaded. As a consequence it is best to keep this resource working all the time by making sure that it always has work, and scheduling it in the best way possible. Therefore it is usually possible to create an optimal schedule on this one resource because of the reduced complexity of the problem. The operations and time before the bottleneck are referred to as a buffer, as is the time following the constraint. The essence of the heuristic is to estimate the time required to have all operations prior to the constraining resource complete before the resource starts, and then do the same for all subsequent operations. The shifting bottleneck method described in the next section is one implementation of this idea. It only produces one schedule, although in case of ties, one could follow each potential path. Also, it is only applicable for problems with multiple resources, although one could use the previous heuristics to schedule the constraint. There are many references that describe Goldratt's ideas in great detail. The preceding explanation should be sufficient for our purposes.

Shifting Bottleneck

One quite successful heuristic for solving the n/m/G/R problem is the shifting bottleneck heuristic. Before we launch into exploring it, we will need a procedure that will help us find lower bounds for whatever measure in which we are interested.

The idea is to 'solve' the problem one machine at a time. The procedure for solving the single-machine problem is frequently referred to as HBT—meaning head, body and tail. The interpretation is that head refers to the work that has to be done on a given job before it can be processed on the machine in question. The body refers to the processing time on the machine in question, while the tail is the remaining processing for a given job after the machine in question. The assumption here is that there is one strong bottleneck that dominates processing—the only one that needs to be scheduled. The implication is that machines preceding and succeeding the bottleneck are plentiful and therefore can process jobs in parallel.

As an example consider a job with technological constraints of M1—M3—M2 and processing times of $p_{i1} = 2$, $p_{i2} = 4$, and $p_{i3} = 3$. Then, when considering M3, the head is 2, the body 3 and the tail 4. M1 refers to all preceding machines, while M3 refers to all succeeding machines.

Once we have these parameters for each job, we are ready to calculate our measure, for example, the smallest feasible maximum completion time C_{max}.

Suppose that we have four jobs, three machines, and the following parameters (Figure 12.9).

As a start, let's evaluate machine 1. From the two tables in Figure 12.9, we derive the head, body, and tail values for each job for this machine (Figure 12.10).

	Processing times					Technological Constraints			
Jobs	**1**	**2**	**3**	**4**	**Totals**	**J1**	**J2**	**J3**	**J4**
p1	5	4	3	1	13	2	1	3	2
p2	5	1	4	3	13	1	2	2	3
p3	2	4	3	3	12	3	3	1	1

Figure 12.9 Longest Tail Example for One Machine, Four Jobs

Machine 1	**Job**	**1**	**2**	**3**	**4**
	head	5	0	7	6
	body	5	4	3	1
	tail	2	5	0	0

Figure 12.10 Machine 1 HBT for Each Job

Our next task is to find the minimum C_{max} for this machine. Note that this is a one-machine problem with non-zero ready times; hence we have no direct algorithm. However, we can apply a heuristic, LT—the longest tail first (due to Schrage, 1970), as follows:

1. Set time t = 0.
2. If there are no jobs available at time t, set time t to the first available job head.
3. Find the job with the largest tail available at time t. Set this job to start at t.
4. Increase t by the processing time of the job selected in step 3 and return to step 2 if not all jobs have been scheduled.

The resulting sequence is optimal if the tail of any job is equal to or shorter than that of any proceeding job (Baker, 2009).

For our example we start with job 2 at t = 0, add 4 to increase t to 4, so the earliest any job can start is 4. When the tail of job 2 is added, C_2 = 9. Job 1 has the next earliest head at 5. When we add the processing time of job 1 to this we get to 10 plus the tail of job 1, which yields C_1=12. The completion of job 1 on the machine is later than the head of either job 3 or job 4. Because each of these has a tail of 0, we can schedule them in either order, getting 2134 or 2143 with a C_{max} of 14 (Figure 12.11). The condition for optimality holds. But remember, this is optimal for machine 1 in isolation; it says nothing about optimality of the complete problem. We can also show that this is optimal either by complete enumeration or integer programming (Chapter 11).

In a similar manner we find the schedules for machines 2 and 3 (Figure 12.12).

In both cases we have actually found the optimum for each machine in isolation. The lower bound must be the largest of these, so we designate machine 2 as the bottleneck and consider its sequence 1243 as solved. The next step is to reevaluate the remaining machines in light of the fixed sequence on machine 2. At this point it is most helpful to introduce the network representation of a general job shop problem.

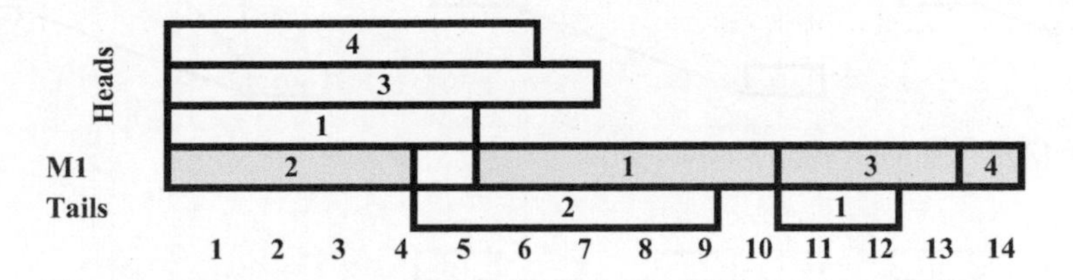

Figure 12.11 Machine 1 Scheduled

Machine 2	Job	1	2	3	4
	head	0	4	3	0
	body	5	1	4	3
	tail	7	4	3	4

Machine 2: 1243 with C_{max} =16

Machine 3	Job	1	2	3	4
	head	10	5	0	3
	body	2	4	3	3
	tail	0	0	7	1

Machine 3: 3421 with C_{max} = 12

Figure 12.12 Machines 2 and 3 Scheduled

In Figure 12.13 we have created a network that proceeds from an initial node at the left through each job in parallel, in its technological constraint sequence, terminating in a node at right. The arrows (referred to as directed arcs because they have a direction) are labeled with the processing time of each operation. This network shows how each job would be processed if it was not restricted in some way by the availability of machines being used by other jobs. The notation in each box is Job, Sequence, and Machine.

Of course, we also need to sequence the jobs on the machines and there are 4! = 24 possible sequences for each of the machines. Consider machine 2 for example and its 24 possible sequences, at first represented by single arrows for two example sequences and then condensed into a set of bidirectional arrows (Figure 12.14).

Figure 12.15 shows all bidirectional arrows (disjunctive arcs). Our objective will be to turn all arrows into one directional or directed ones and remove any that could lead to cycles (any path that could lead back to a node that already has an in-arc) in our network.

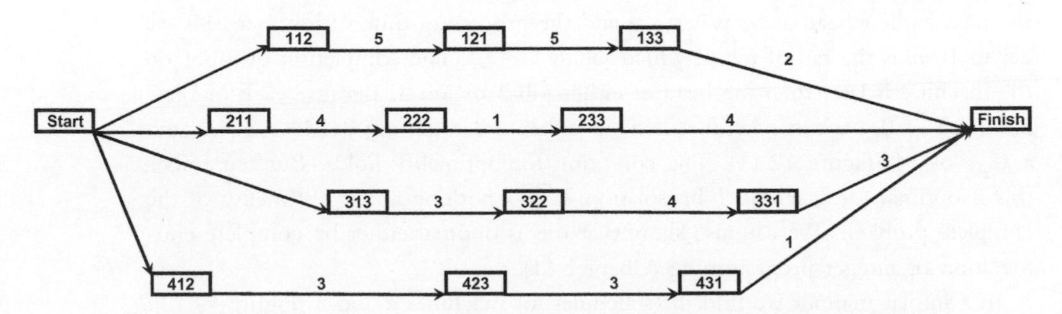

Figure 12.13 Network Representation of the Four-Job, Three-Machine Example

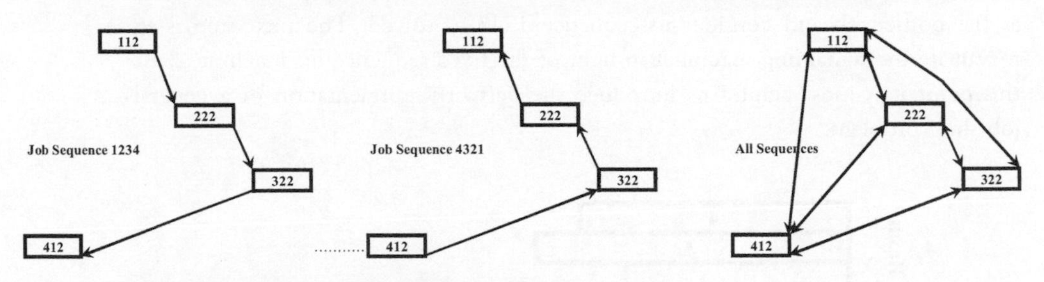

Figure 12.14 Sequences on Machine 2

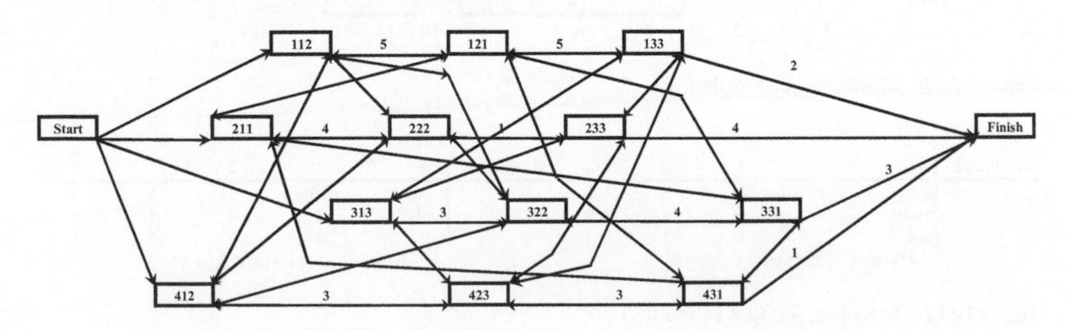

Figure 12.15 All Disjunctive Arrows

If we use an active schedule process (see Chapter 7), we may get a schedule such as (of course many other feasible ones are possible—this one may or may not be optimal):

M1: 2431; M2: 4321; M3: 3421 shown in the Gantt chart of Figure 12.16.

We can prune our network to represent it (note that there are three paths—shaded areas—that give us our longest path of 20 in Figure 12.16) in Figure 12.17 where the longest paths are indicated by dashed arrows.

We can now return to our shifting bottleneck procedure. Where we left it, we had arrived at a sequence of 1243 for machine 2. Because we now know how to represent our schedules in a network, we show this result below (Figure 12.18).

Our next step is to evaluate machines 1 and 3 with the sequence on machine 2 in place. To accomplish this evaluation, it is helpful to indicate the times that are now constrained on our network due machine 2's schedule (shown in underlined

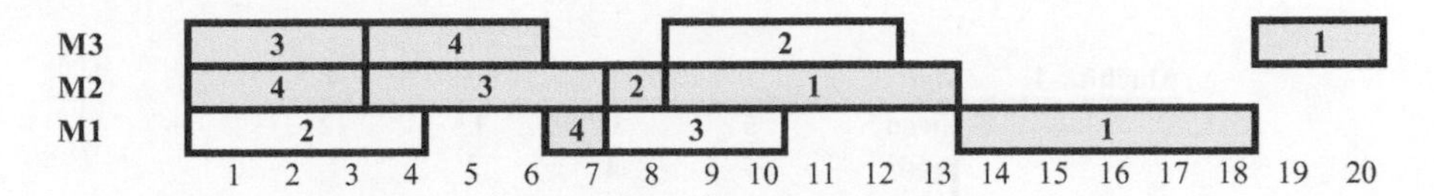

Figure 12.16 Gantt Chart for the Example Active Schedule

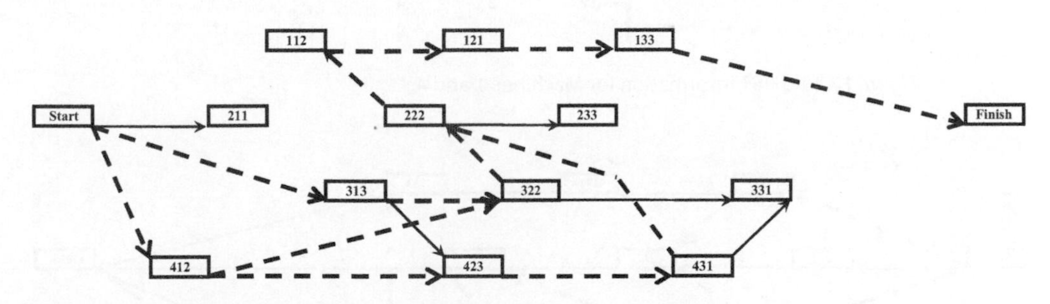

Figure 12.17 Network Representation of the Example Active Schedule

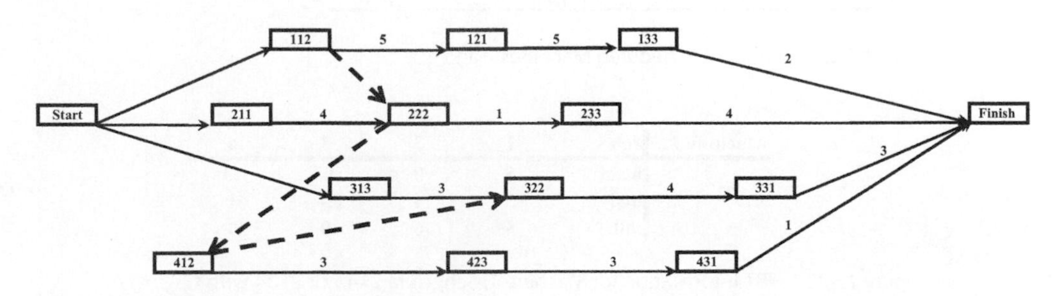

Figure 12.18 Resolved Schedule for Machine 2

italics above the nodes of the network affecting machines 1 and 3 in Figure 12.19). The resulting HBT information for these machines is shown in Figure 12.20.

From these, using HeadBodyTail, we get M1: 2143 with C_{max} = 16 and M3: 3241 also with C_{max} = 16. So we pick M3 arbitrarily as our next bottleneck and we fix its sequence. The effect is shown in Figure 12.21. Figure 12.22, shows the HBT for the remaining machine 1.

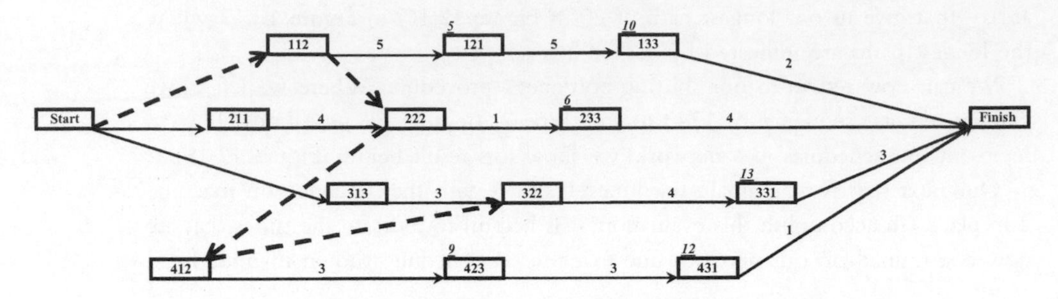

Figure 12.19 Effect of Scheduled Machine 2 on the Other Two Machines

Machine 1	Job	1	2	3	4
	head	5	0	13	12
	body	5	4	3	1
	tail	2	6	0	0

Machine 3	Job	1	2	3	4
	head	10	6	0	9
	body	2	4	3	3
	tail	0	0	13	1

Figure 12.20 HBT Information for Machines 1 and 4

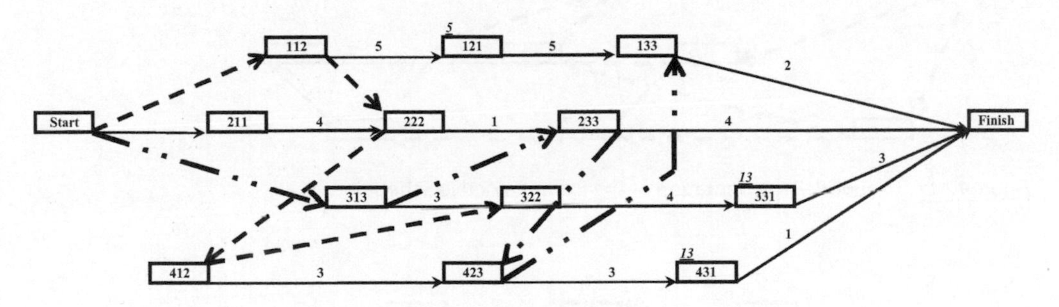

Figure 12.21 Effect of Scheduled Machines 2 and 3

Machine 1	Job	1	2	3	4
	head	5	0	13	13
	body	5	4	3	1
	tail	5	6	0	0

Figure 12.22 HBT Information for Machine 1, Schedule 2143 or 2134 with C_{max} = 17

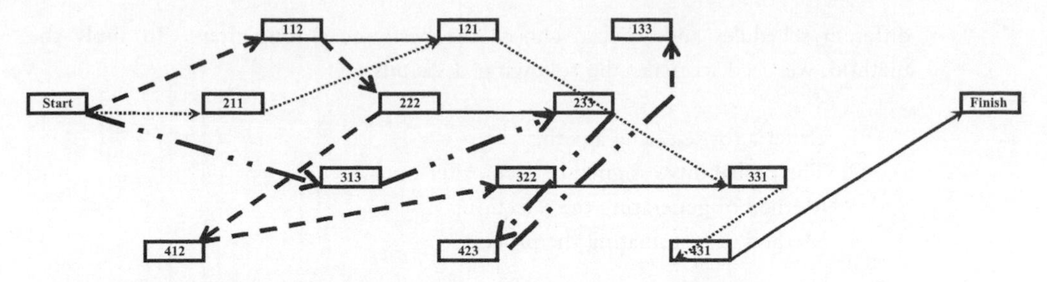

Figure 12.23 Final Resolved Network

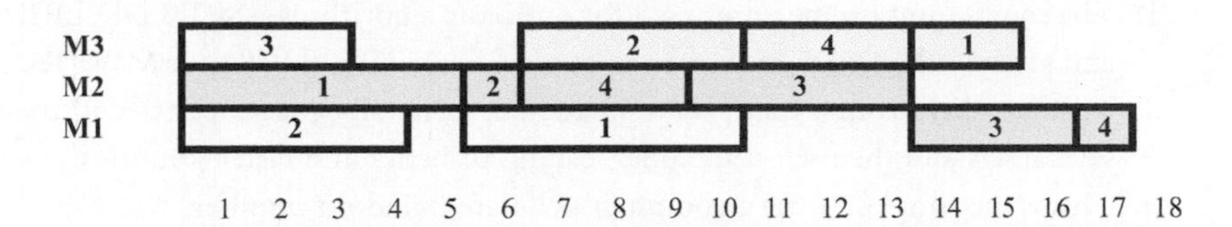

Figure 12.24 Gantt Chart for the Shifting Bottleneck Solution

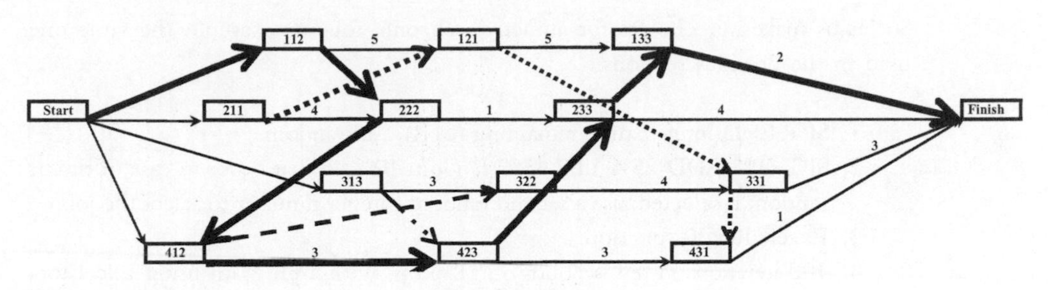

Figure 12.25 Final Network—with the Critical (Longest) Path

Our final network is shown in Figure 12.23.

It remains to draw our Gantt chart and show the longest path in our network (Figures 12.24 and 12.25).

This result is three units better than what we got with the one active schedule that we generated. Usually we do not know what the optimum is, but in this case it is optimal. Why?

Monte Carlo Methods

The Monte Carlo method has been around for a long time. It takes its name from the casino in Monte Carlo known worldwide for its gambling. The essence is that any outcome is probabilistic, such as drawing a specific card from a deck, or a specific number coming up in roulette. As we saw in Chapter 7 and in the active schedule of the shifting bottleneck section, there are many potential paths in generating a schedule. The Monte Carlo method probabilistically selects a choice among potential paths at each step. Repeating the process many times will produce many different schedules and we can choose the best one among them. To apply the method, we need to make the following decisions:

1. Criteria for selecting a path;
2. The probability assigned to each criterion;
3. Method of generating the selection;
4. Method of terminating the process.

The generally available choices for each of these are as follows:

1. The most common measures used for our basic algorithms—SPT, EDD, LPT, and several others such as least or most work remaining (LWKR or MWKR).
2. This is rather arbitrary and very situational. Generally, practitioners will try several sets and then select the ones leading to better and faster solutions.
3. This is the simplest—use a computer generated random number.
4. Some of the choices here are length of time, number of schedules, or schedules elapsed without an improvement.

So let us make our choices for an active schedule for our example, the same one used in the previous sections.

1. SPT, EDD, longest time remaining (LTR), and random.
2. SPT 50%, EDD 25%, LTR 15%, random 10%, in that order in case of ties. If random is selected, use a second random integer number to select the job.
3. Excel's Rand() function.
4. 100 iterations (a few seconds on a laptop) with a program using Excel formulas and some VBA (Visual Basic) programming.

Before we use the program, let's go through a manual procedure to clarify the process. We begin by creating a table (Figure 12.26) for the first of the 12 stages (three machines, four jobs). Each operation is indicated by (j, m) where j is the job and m is the machine. This is determined by the current state (in this case, the initial state) of the schedule and the technological constraints.

The active schedule rule says to select the machine with the earliest finishing job. In this case it is either machine 2 or 3, and the choice is arbitrary because they are on different machines (we will return to the machine not chosen at a later

Stage 1	Operation	Start	Duration	Finish	Remaining Operations	Remaining Time	Select Machine	Select Job	Reason
	(1,2)	0	5	5	3	12			
	(2,1)	0	4	4	3	9			
	(3,3)	0	3	3	3	10	*	*	Rand 0.54 LTR
	(4,2)	0	3	3	3	7			

Figure 12.26 Initial Stage of the Monte Carlo Procedure for an Active Schedule

stage without impeding it). Suppose we say that a random number lower than 0.5 chooses machine 2, otherwise machine 3 and we obtain a random number of 0.6. We assign job 3 to machine 3. If you are doing this manually, it is best to build a Gantt chart as you go along (Figure 12.27). Also, mark the selected operation in both the table of technological constraints and durations of operations (Figure 12.9).

At stage 2 we have the scenario in Figure 12.28. Job 3 is now ready to go on machine 2 at time 3, its finishing time from the last stage. The other jobs are not affected.

The earliest completion is on machine 2. However, job 3 is not eligible, as it does not start before the completion of job 4. So we are left with a choice between jobs 1 and 4. We need to select a random number: 0.7. So we select by longest remaining time, which is job 1. We now have the Gantt chart in Figure 12.29. The remaining stages and the final Gantt chart are in Figure 12.30 and Figure 12.31.

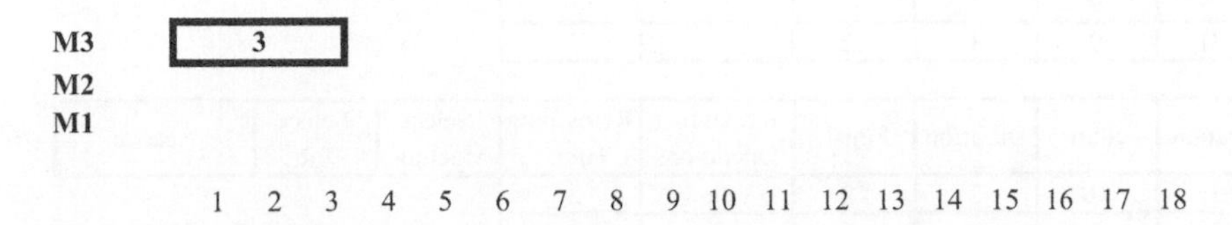

Figure 12.27 Starting the Gantt Chart for the Monte Carlo Procedure

Stage 2	Operation	Start	Duration	Finish	Remaining Operations	Remaining Time	Select Machine	Select Job	Reason
	(1,2)	0	5	5	3	12		*	Rand 0.7 LTR
	(2,1)	0	4	4	3	9			
	(3,2)	3	4	7	2	7			
	(4,2)	0	3	3	3	7	*		

Figure 12.28 Second Stage of the Monte Carlo Procedure for an Active Schedule

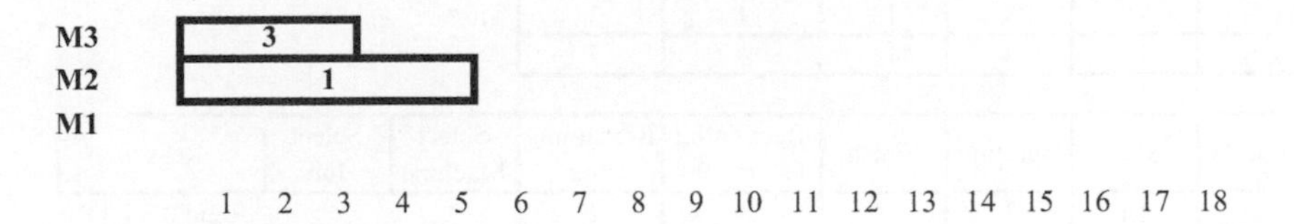

Figure 12.29 Gantt Chart for Stage 2 for the Monte Carlo Procedure

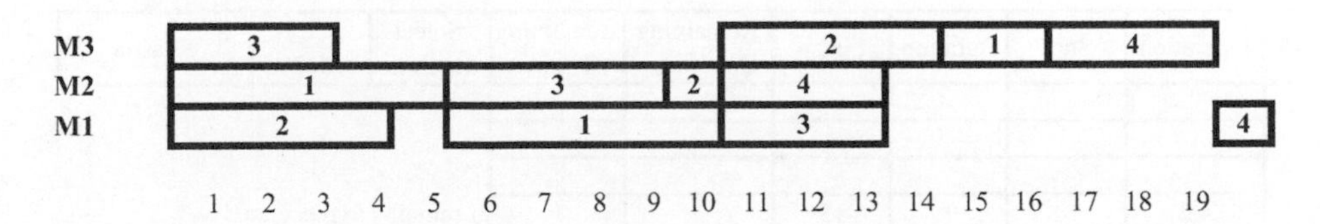

Figure 12.30 Complete Gantt Chart for the Monte Carlo Procedure

Stage 3	Operation	Start	Duration	Finish	Remaining Operations	Remaining Time	Select Machine	Select Job	Reason
	(1,1)	5	5	10	2	7			
	(2,1)	0	4	4	3	9	*	*	only
	(3,2)	5	4	9	2	7			
	(4,2)	5	3	8	3	7			

Stage 4	Operation	Start	Duration	Finish	Remaining Operations	Remaining Time	Select Machine	Select Job	Reason
	(1,1)	5	5	10	2	7			
	(2,2)	5	1	6	2	5	*		
	(3,2)	5	4	9	2	7		*	Rand .99, 3
	(4,2)	5	3	8	3	7			

Stage 5	Operation	Start	Duration	Finish	Remaining Operations	Remaining Time	Select Machine	Select Job	Reason
	(1,1)	5	5	10	2	7	*	*	only
	(2,2)	9	1	10	2	5			
	(3,1)	9	3	12	1	3			
	(4,2)	9	3	12	3	7			

Stage 6	Operation	Start	Duration	Finish	Remaining Operations	Remaining Time	Select Machine	Select Job	Reason
	(1,3)	10	2	12	1	2			
	(2,2)	9	1	10	2	5	*	*	Rand 0.29 SPT
	(3,1)	10	3	13	1	3			
	(4,2)	9	3	12	3	7			

Stage 7	Operation	Start	Duration	Finish	Remaining Operations	Remaining Time	Select Machine	Select Job	Reason
	(1,3)	10	2	12	1	2	*		
	(2,3)	10	4	14	1	4		*	Rand 0.55 LTR
	(3,1)	10	3	13	1	3			
	(4,2)	10	3	13	3	7			

Stage 8	Operation	Start	Duration	Finish	Remaining Operations	Remaining Time	Select Machine	Select Job	Reason
	(1,3)	14	2	16	1	2			
	(3,1)	10	3	13	1	3	*	*	Only
	(4,2)	10	3	13	3	7			

Stage 9	Operation	Start	Duration	Finish	Remaining Operations	Remaining Time	Select Machine	Select Job	Reason
	(1,3)	14	2	16	1	2	*	*	Rand 0.8 LTR
	(4,2)	10	3	13	3	7			

Stage 10	Operation	Start	Duration	Finish	Remaining Operations	Remaining Time	Select Machine	Select Job	Reason
	(4,2)	10	3	13	3	7	It remains to placejob 4		

Figure 12.31 Stages 3–10 of the Monte Carlo Procedure for an Active Schedule

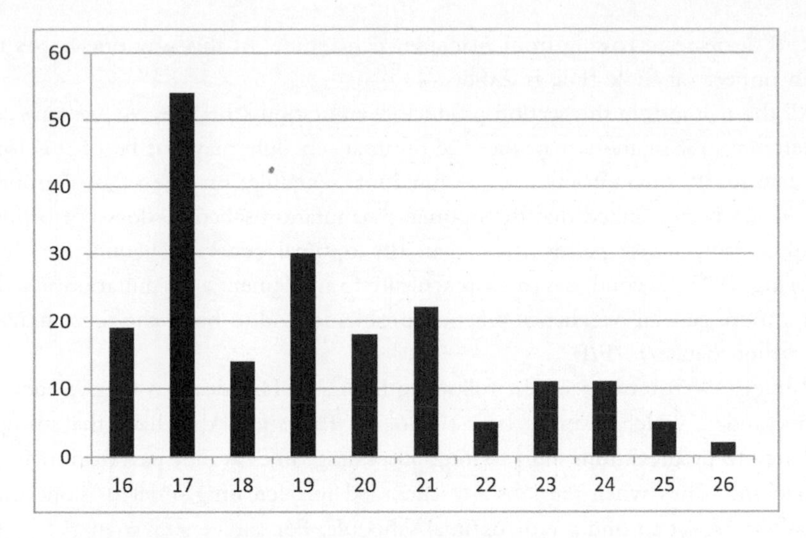

Figure 12.32 Distribution of C_{max} of the Active Schedules

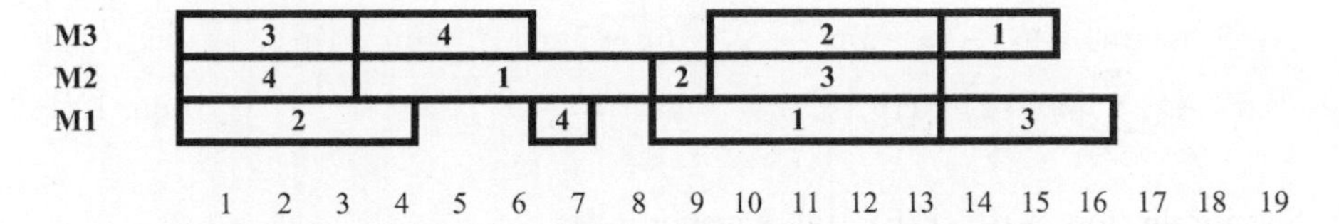

Figure 12.33 One Optimal C_{max} Active Schedule

Note that this result (Figure 12.32) is different from the previous two results. One performance of the Monte Carlo simulation found the optimum 16 times. In fact, there are 193 active schedules if you follow each potential path. If you do that you will find that the optimum C_{max} is 16. The distribution of the active schedules is shown in Figure 12.32 (Gantt chart in Figure 12.33).

Flow Shop Scheduling Heuristics

Now we turn to a group of methods designed specifically to tackle flow shop scheduling and so quite inapplicable to other classes of problem. They are all based upon our physical intuition of the structure of a near optimal schedule in a flow shop with maximum flow time as the performance measure. Throughout this section we assume that all the ready times are zero.

Remember that in motivating Johnson's algorithm for the $n/2/F/F_{max}$ problem we argued that the jobs that were placed near the beginning of the schedule should have short processing times on M1. Idle time on M1 is then minimized. Similarly, jobs that were placed near the end of the schedule should have short M2 processing times in order to minimize M2 idle time. The methods below carry over this idea to the general $n/m/F/F_{max}$ problem. It is intuitively reasonable that jobs placed early in the schedule should have processing times that tend to increase from machine to machine, whereas those toward the end of the schedule should have times that tend to decrease in passing from machine to machine. In this way one hopes that much unnecessary idle time is avoided.

All the methods in this section produce permutation schedules. As we have seen if there are four or more machines the optimal schedule may not be of this form. We may justify this restriction to permutation schedules on one of two grounds. First, it has been claimed that the optimal permutation schedule does not produce an appreciably worse performance than the optimal general schedule (see Dannenbring, 1977). Second, it is easier practically to implement a permutation schedule than a more general one; hence practical problems tend to be of the form $n/m/P/F_{max}$ rather than $n/m/F/F_{max}$.

Palmer (1965) introduced the following heuristic. His idea was to give each job a slope index, which gives its largest value to those jobs that have the strongest tendency to progress from short to long processing times as they pass from machine to machine. Thus when the jobs are scheduled in decreasing order of slope index we might expect to find a near optimal schedule. For the general $n/m/F/F_{max}$ this slope index for J_i is

$$\begin{aligned} S_i &= -(m-1)p_{i1} - (m-3)p_{i2} - \ldots + (m-3)p_{i(m-1)} + (m-1)p_{im} \\ &= -\sum_{j=1}^{m}\left[m-(2j-1)p_{ij}\right] \end{aligned} \qquad \text{Equ. 12.1}$$

Specifically, for an $n/7/F/F_{max}$ this equation gives

$$S_i = -6p_{i1} - 4p_{i2} - 2p_{i3} + 0p_{i4} + 2p_{i5} + 4p_{i6} + 6p_{i7}$$

It may be seen that when the process times tend to increase from machine to machine, the early terms in this sum are small and negative while the later terms are large and positive. This index does have the suggested property.

As an example we apply Palmer's algorithm to the $4/3/F/F_{max}$ problem used in Chapter 10 The data are repeated in Figure 12.34.

The slope indices:

$$S_1 = -2 \times 1 + 0 \times 8 + 2 \times 4 = 6$$
$$S_2 = -2 \times 2 + 0 \times 4 + 2 \times 5 = 6$$
$$S_3 = -2 \times 6 + 0 \times 2 + 2 \times 8 = 4$$
$$S_4 = -2 \times 3 + 0 \times 9 + 2 \times 2 = -2$$

Job	Processing Times		
J_i	a_i	b_i	c_i
1	1	8	4
2	2	4	5
3	6	2	8
4	3	9	2

Figure 12.34 Data from the Branch and Bound Example of Chapter 10

The algorithm suggests the processing sequence (1, 2, 3, 4) or (2, 1, 3, 4). To decide which to use we would evaluate F_{max} for each of these and choose the better. In fact, both schedules are optimal and share $F_{max} = 28$. Here we are lucky in that optimal solutions are found. Dannenbring (1977) has found empirically for a sample of 1280 small problems ($n \leq 6$, $m \leq 10$) that Palmer's heuristic methods finds the optimum in about 30% of the cases.

Campbell, Dudek, and Smith (1970) use a multiple application of Johnson's two machine algorithm to try to obtain a good schedule. Essentially, they create (m – 1) scheduling problems; use Johnson's algorithm on each of these; and pick the best of the resulting (m – 1) schedules. To be precise, the kth problem (k = 1, 2, . . ., (m – 1)) is formed as follows. The constructed processing time on the 1st machine for the ith job is $a_i^{(k)} = \sum_{j=1}^{k} p_{ij}$, that is it is the sum of processing times for the ith job on the first k actual machines. Similarly the constructed time on the second machine is the sum of processing times for the ith job on the last k actual machines: $b_i^{(k)} = \sum_{j=-k+1}^{k} p_{ij}$

Note that for m = 3, k = 2, this corresponds to an application of Johnson's algorithm for the special case of the $n/3/F/F_{max}$ problem. Campbell, Dudek, and Smith's (CDS) algorithm may be seen as an intuitive generalization of this. Note also that this multiple application of Johnson's Algorithm will tend to yield schedules of the type expected by our intuitive arguments above.

Applying this algorithm to our example we construct 2 two-machine problems corresponding to k = 1 and k = 2 (Figure 12.35).

Applying Johnson's algorithm gives the processing sequence (1, 2, 3, 4) for the first constructed problem and (2, 3, 1,4) for the second. The first we know has $F_{max} = 28$; the second has $F_{max} = 29$. So the method would choose the sequence (1, 2, 3, 4) and as Palmer's did, achieve optimality in this case. Dannenbring's empirical results for the 1,280 small problems showed that this algorithm achieves optimality roughly 55% of the time.

Dannenbring (1977) has himself designed a heuristic method, which attempts to combine the advantages of Palmer's and Campbell, Dudek, and Smith's. His idea is to construct just one two-machine problem on which to apply Johnson's algorithm, but with the constructed processing times reflecting the same behavior as Palmer's slope index. The constructed time for the first machine is

$$a_i = mp_{i1} + (m-1)p_{i2} + \ldots + 1p_{im} = \sum_{j=1}^{m}(m-j+1)p_{ij} \qquad \text{Equ. 12.3}$$

Job				Constructed Problem k = 1		Constructed Problem k = 2	
Ji	p_{i1}	p_{i2}	p_{i3}	$a_{i(1)} = p_{i1}$	$b_{i(1)} = p_{i2}$	$a_{i(2)} = p_{i1}$	$b_{i(2)} = p_{i2}$
1	1	8	4	1	4	9	12
2	2	4	5	2	5	6	9
3	6	2	8	6	8	8	10
4	3	9	2	3	2	12	11

Figure 12.35 Calculating CDS Sequences

and that for the second machine is

$$b_i = p_{i1} + 2p_{i2} + \ldots + 1p_{im} = \sum_{j=1}^{m} jp_{ij} \qquad \text{Equ. 12.4}$$

Thus a job's constructed time a_i is relatively short if it tends to have short actual processing times on the early machines. Similarly b_i is relatively short if J_j tends to have short actual processing times on the later machines.

Applying this method to our example in Figure 12.36:

	Processing times			Constructed Processing Times	
Ji	pi1	pi2	pi3	$a_i = 3p_{i1} + 2p_{i2} + p_{i3}$	$b_i = p_{i1} + 2p_{i2} + 3p_{i3}$
1	1	8	4	23	29
2	2	4	5	19	25
3	6	2	8	30	34
4	3	9	2	29	27

Figure 12.36 Dannenbring Coefficients

The application of Johnson's algorithm gives the processing sequence (2, 1, 3, 4), which has F_{max} = 28 and again is optimal. Dannenbring's empirical results show that for small problems this method finds the optimal schedule in about 35% of all cases.

Dannenbring compared these three algorithms and some others not discussed here on many problems, both large and small. Overall he found that Campbell, Dudek, and Smith's algorithm performed much better than the other two. The most important result of Dannenbring's study is that he found that all these methods could be dramatically improved by using their output as initial seeds in a neighborhood search routine, even if one limited the search to just one cycle. This is perhaps the most promising direction to investigate in the design of flow shop heuristics. Dannenbring's experiment was repeated by this author with very similar results for 100 problems ranging from m = 4 to m = 6 and processing times assigned randomly between 1 and 12 to four to eight jobs: Palmer 19%, CDS 36%, and Dannenbring 49%. In addition, we found that the average F_{max} obtained was 4% over the optimum for Palmer, Dannenbring on 3%, and CDS on 1%. It is somewhat ironic that Dannenbring should only prove his own method to be second best. These heuristics have stood the passage of time and are still the best ones available.

13 PARALLEL MACHINES AND WORST CASE BOUNDS

Introduction

The last chapter explored the idea of heuristic methods. We referred to some empirical studies that showed that in the majority of cases these 'sensible' rules generate reasonably good schedules. However, they cannot guarantee optimality and there is always a nagging doubt that in some unfortunate cases they may lead to very poor results. Our first objective in this chapter is to consider worst case bounds on the performance of certain heuristic algorithms, i.e., to discover how poorly they may perform in the worst of possible circumstances.

We must use heuristic methods because many scheduling problems are NP-hard (see Appendix C), so there are unlikely to be polynomial time algorithms for finding optimal schedules. We have stopped seeking the best possible solution, and agreed to accept one that is not too poor. Once we have introduced the concept of a worst case bound on the sub-optimality of heuristically generated schedules, it is natural that we ask how well can we do in polynomial time. It does not make sense to avoid an exponential time optimization algorithm by using an approximate method that takes as long. Given that we limit ourselves to polynomial time heuristic algorithms, what sort of performance can we guarantee? We shall find that the theory of NP-completeness again presents us with a very pessimistic outlook.

Our main assumption is that all the numerical quantities that define an instance are integers. This is not a problem—we can always multiply all quantities with a sufficiently large number to obtain all integers. Note that the class of scheduling problems considered in the next section breaks many of the assumptions of Chapter 6.

Scheduling Independent Jobs on Identical Machines

For this section there is some advantage to be gained by considering an entirely distinct class of scheduling problems from those that we have discussed before. This has the advantage that for this one class we can find heuristic algorithms that exhibit each of the different possible patterns of worst-case behavior that we want to discuss. We consider the problem of scheduling n independent jobs $\{J_1, J_2, \ldots, J_n\}$ on m identical machines $\{M_1, M_2, \ldots, M_m\}$. Each job J_i requires only one processing operation and that operation may occur on any of the m machines. Because the machines are identical, the processing time, p_i of job J_i is the same on any of the machines. The jobs may be processed in any order and are available at time zero (all $r_i = 0$). No machine may process more than one job at a time. Our objective is to minimize C_{max}, the time that the last job is completed.

As an example, consider a college lab with m computers all with identical processing speeds. At the beginning of a day the lab supervisor has requests from n students and her task is to assign them to the computers so that all students have completed their work and the lab is cleared as soon as possible. Of course, in other examples the machines (here the computers) do have different speeds (non-identical machines), some jobs have to be done before others (precedence constraints), and the jobs do not all arrive together (non-zero ready times or stochastic problems). For the time being we ignore such complications. For $m > 1$ this problem is NP-complete (has no polynomial complexity algorithm), so it does make sense to consider heuristic methods. (When $m = 1$ this problem becomes the trivial $n/1//C_{max}$ problem, for which C_{max} is a constant independent of the schedule.)

Define an instance, I, of the problem as a specification of the data n, m, p_1, p_2, . . ., p_n. We define OPT(I) to be the maximum completion time of all the jobs if an optimal schedule is used for instance I. A(I) is the maximum completion time if a particular heuristic algorithm A is used for instance I. We look for bounds on the relative performance. $R_A(I) = A(I)/OPT(I)$. Note that $RA(I) \geq 1$ for all A and for all I. The better the schedule found by A the nearer $R_A(I)$ is to 1.

The first heuristic method that we consider is the list scheduling algorithm. This lists the jobs in any order and then assigns them in this order to the machines as they become free. Note that the list of jobs is in a completely arbitrary order. The idea can most easily be seen through an example. Given six jobs and three machines and the processing times:

$$p_1 = 8,\ p_2 = 2,\ p_3 = 2,\ p_4 = 8,\ p_5 = 1,\ p_6 = 4$$

Suppose that we use the randomly generated list (J_4, J_2, J_5, J_3, J_1, J_6). At $t = 0$ all machines are available for processing so we assign J_4 to M_1, J_2 to M_2, and J_5 to M_3. Because more than one machine was available, we have made the conventional arbitrary choice of assigning jobs to machines in order of increasing machine index. The first machine to finish its task is M_3 at $t = 1$, so we assign J_3 to this next. Again M_3 finishes before the others, so we assign J_1 to it. Next at $t = 6$, M_2 completes and is assigned J_6. All processing eventually completes at $t = 11$. This is illustrated in the Gantt diagram in Figure 13.1.

Theorem 13.1 (Graham, 1969) If A is the list scheduling algorithm, then for all instances I:

$$R_A(I) \leq 2-1/m. \qquad \text{Equ. 13.1}$$

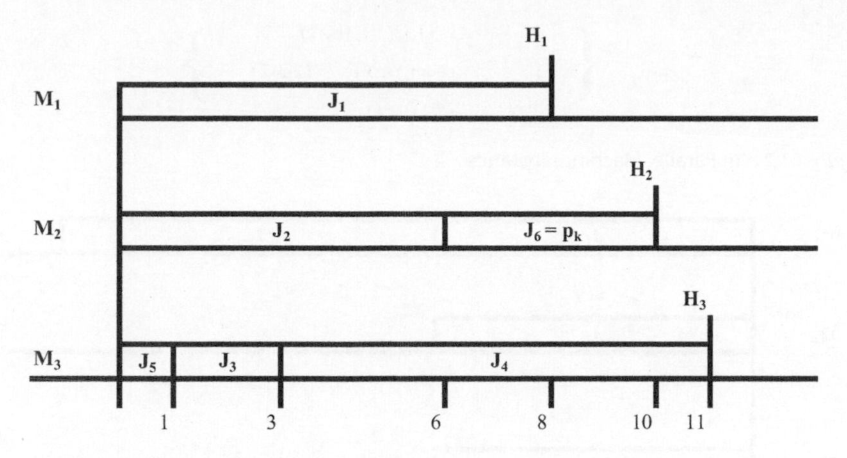

Figure 13.1 Gantt Diagram for the List Scheduling Example

Proof First we find two lower bounds on OPT(I). We cannot do better than the longest processing time among the jobs:

$$\mathrm{OPT}(\mathrm{I}) \geq \max_{i=1}^{n}\{p_i\} \qquad \text{Equ. 13.2}$$

At best the processing is divided equally between the machines. Then:

$$\mathrm{OPT}(\mathrm{I}) \geq \frac{1}{m}\sum_{i=1}^{n} p_i \qquad \text{Equ. 13.3}$$

Let H_j be the time that processing finally halts on machine M_j. Let J_k be a job that completes at A(I), that is J_k completes last. Because the list scheduling algorithm always assigns the next job to a machine as soon as it finishes processing its current job, no machine can cease processing while there are any unassigned jobs in the list.

Therefore none can stop before the processing of J_k starts:

$$H_j \geq A(I) - p_k$$

We also know that at least one machine halts at A(I). So

$$\sum_{i=1}^{n} p_i = \sum_{j=1}^{m} H_j \geq (m-1)(A(I) - p_k) + A(I)$$

Then by Equations 13.1 and 13.2, $A(I) \leq \frac{1}{m}\sum_{i=1}^{n} p_i + \frac{(m-1)}{m} p_k \leq OPT(I) + \frac{(m-1)}{m}$ $OPT(I)$.

Therefore $R_A(I) = \frac{A(I)}{OPT(I)} \leq 2 - \frac{1}{m}$. We can show that this bound is tight, that there exists an instance and a list such that $R_A(I) = \frac{A(I)}{OPT(I)} \leq 2 - \frac{1}{m}$.

$$p_i = \left\{ \begin{array}{ll} (m-1) & i = 1, 2, \ldots, (m-1) \\ 1 & i = m, (m+1), \ldots (2m-2) \\ m & i = (2m-1) \end{array} \right\}$$

Figure 13.2 m Parallel Machines Instance

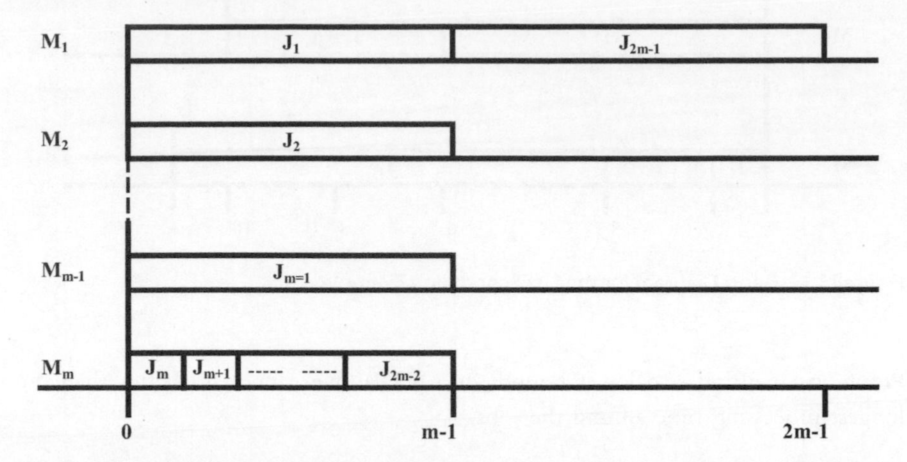

Figure 13.3a List Schedule: A(I) = 2m – 1

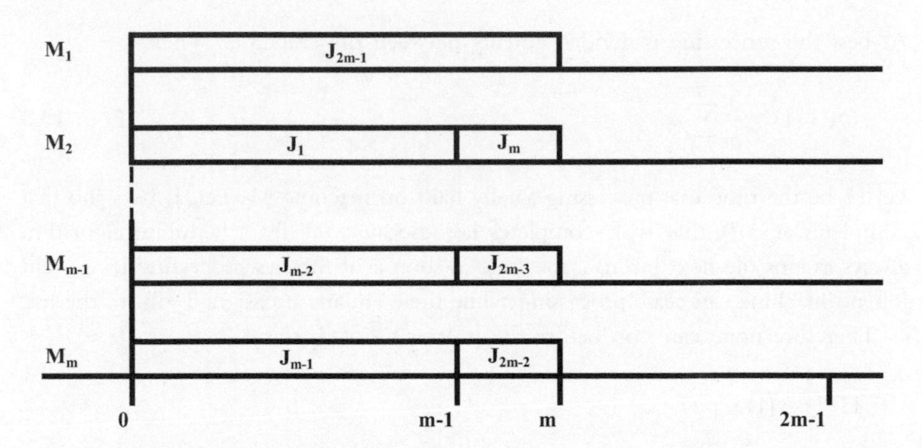

Figure 13.3b Optimal Schedule: A(I) = m

Consider an instance with m machines and (2m – 1) jobs. Let the processing times be as shown in Figure 13.2.

If we use the natural order list $(J_1, J_2, \ldots, J_{(2m-1)})$ in the list scheduling algorithm, we obtain the schedule shown in Figure 13.3a. This has A(I) = (2m – 1). If on the other hand we use the list $(J_{2m-1}, J_1, J_2, \ldots, J_{(2m-1)})$ then we obtain the schedule shown in Figure 13.3b. This is clearly optimal, because the processing is divided evenly between the machines. Thus OPT(I) = m. Hence here $R_A(I) = A(I)/OPT(I) = (2m - 1)/m = 2 - 1/m$, and we have an instance and a list that attains the bound in Theorem 13.1. We call the bound provided by Theorem 13.1 a performance guarantee and we call this style of analysis worst case analysis.

As an example we apply this theorem to the example of the computer lab mentioned earlier. Suppose that there are four computers with identical speeds and that the supervisor assigns work to them by the list scheduling algorithm with an arbitrarily ordered list, say the order in which the students signed up. Then the overall time to complete the work could be up to 2 – 1/4 = 1.75 as long as it would be if an optimal assignment were used. In other words, Theorem 13.1 provides a guarantee that in the worst possible case assigning work in the order of the signings can lead to a delay in completion relative to the optimal of no more than 75%. Are there are other algorithms that give better performance guarantees than the list scheduling algorithm? There are. A natural improvement is to order the jobs in the list in more sensible fashion: Looking at Figure 13.3a suggests that the poor performance of the algorithm in the first case is due to the longest job occurring last in the list. Thus it is natural to consider the longest processing time algorithm. This modifies the list scheduling algorithm by demanding that the jobs are ordered in non-increasing processing time in the list. For instance, if we consider the six-job machine example, we see that a non-increasing list of jobs is (J_1, J_4, J_2, J_6, J_3, J_5).

This gives the schedule shown in Figure 13.4. Because all the p_i are integral and because $\frac{1}{5}\sum_{i=1}^{6} p_i = 9\frac{2}{3}$, this schedule must be optimal. Unfortunately, the longest processing time schedule does not always find the optimal schedule and it can be proven (Graham, 1969) that the following performance is guaranteed:

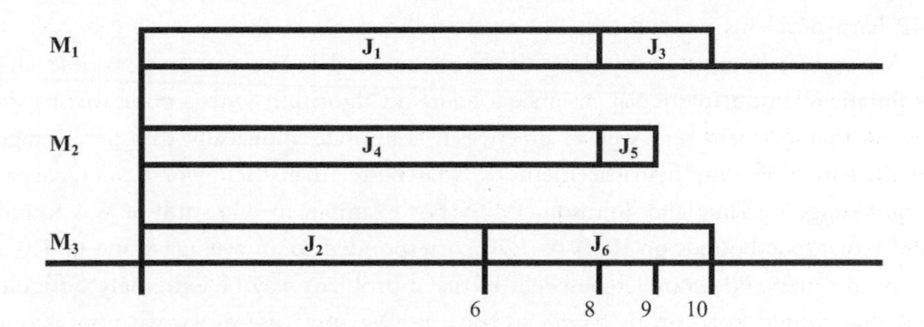

Figure 13.4 Gantt Diagram for the LPT Scheduling Example

Theorem 13.2 If A is the longest processing time algorithm, then

$$R_A(I) = \frac{A(I)}{OPT(I)} \leq \frac{4}{3} - \frac{1}{3m} \quad \text{for all instances I}$$

If we return to the lab example, we may see the improvement that this algorithm brings. Suppose that the supervisor now assigns students to the terminals in order of non-increasing processing time; then the work will be completed within (4/3 – 1/12) = 1 1/4 times that resulting from the optimal assignment.

The preceding development illustrates one stage in a cyclical process by which worst case analysis can lead to better and better heuristic algorithms. Initially an apparently sensible algorithm is proposed, here the list scheduling algorithm. Guarantees are found upon its performance (Theorem 13.1) and, more importantly for algorithm design and improvement, examples that exhibit weaknesses are sought. Improvements are made to the algorithm to counter these; here we developed the longest processing time algorithm. Performance guarantees are sought upon this new algorithm to ensure that, in the sense of the bounds, there has indeed been an improvement. Given that there has, the new algorithm becomes a candidate for improvement and the process cycles onward.

Worst Case Analysis and Expected Performance

The theory of NP-completeness provided us with a very pessimistic outlook. Already we have learned that for many scheduling problems there is little chance of developing polynomial time-optimizing algorithms.

But let us try to be more positive. As stated earlier, practical scheduling problems cannot be left unsolved; some processing sequence must be used and, depending on that sequence, costs will be incurred. Surely we can do better than simply pick a schedule at random. So suppose that we are faced with a large NP-hard problem. What should we do?

First, we should remember that the theory of NP-completeness applies to classes of problems. We are faced with a particular instance. Branch and bound or some other form of implicit enumeration may solve this instance quickly. Thus despite all our fears we may find an optimal schedule. Nevertheless, for the majority of NP hard problems we will fail. We must try heuristic methods.

Worst case bounds are by definition only attained in the worst of possible circumstances. In our particular instance a heuristic algorithm with a poor worst case bound may perform very well in practice. It is known empirically that the average performance of some heuristic methods is far better than their worst case analysis might suggest (Garey and Johnson, 1979). For example, in one study it was found that worst case bounds on $R_A(I)$ of 1.70 corresponded to an average value of 1.07.

In summary, NP-completeness tells us that a problem may be extremely difficult, but that should not stop us trying to solve it. We must use every weapon at our disposal and do the best that we can.

Scheduling Independent Jobs on Uniform Machines

Uniform machines are defined as machines that perform the same work, but at different speeds. One example is an old milling machine compared to an automated machining cell. Another more immediate example is two students taking the same exam; one is more than likely faster at it than another, even if both know exactly what needs to be done. When we have a number of machines with equal abilities but different speeds available, we would like to know how to assign jobs to them to minimize the maximum completion time. We of course do require to know the processing time of each job on each machine. In practice we would probably just assign the processing in the same ratio for all jobs to the slower machines, having determined the time on the fastest machine.

Because there is no known polynomial algorithm for such a problem, we again resort to heuristics. This one starts with the idea that if we could we would want to process all jobs on the fastest machine, but that we can reduce that C_{max} by moving some jobs to the slower machines. Also, we can still resort to the better of the two heuristics for equivalent machines, that is the longest processing time procedure.

Heuristic for uniform machines:

1. Order all jobs in longest processing time order on the fastest machine.
2. Remove the first job from the fastest machine to the next fastest machine.
3. Continue this process until a move results in a higher C_{max}, in which case undo the move.
4. Repeat until you have used some time on all the machines. Once you are working on the third machine, it is not defined whether to move jobs from the fastest or second fastest machine.

Example The processing times on three machines for ten jobs are given in Figure 13.5. Determine a good solution to minimizing the maximum completion time.

The solution is obtained using Excel's lookup functions to find the processing time of each job on each machine from the given data. First organize all jobs on all three machines in LPT order. If you substitute 0 for the number of any job on a machine, its processing time also becomes zero and it is effectively eliminated from that machine.

Jobs	1	2	3	4	5	6	7	8	9	10	11	12	13
Machine 1	8	5	6	9	9	5	7	7	9	4	10	3	7
Machine 2	9	5	7	10	10	6	8	8	10	4	11	6	7
Machine 3	10	6	9	11	11	7	9	9	11	5	12	8	8

Figure 13.5 Data for the Uniform Machine Example

	Job	compl	Job	compl	Jcb	compl	Job	compl	Job	compl	Job	compl	Job	compl	Job	compl	Job	compl	Job	compl	Job	compl	Job	compl	Job	compl
M2	0	0	0	0	0	0	0	0	0	0	0	0	0	0	7	7	13	14	2	19	6	24	11	34	12	37
M1	0	0	0	0	0	0	9	10	8	18	1	27	3	34												
M3	10	5	4	16	5	27																				

Figure 13.6 Solution of the Uniform Machine Example

14 RELAXATION OF ASSUMPTIONS

Sequence Dependent Set Ups (SDSU)

When we first treated one-machine problems, we assumed that any setup times for a job were included in the processing time and hence constant with no impact on optimizing a measure. However, when changing from one job to another, the setup time may be dependent on the previous job. One example is the exchange of molds and plastic molding powders in injection molding. The time required to cool and heat the molds and to purge the machine of the previous powder is dependent on the difference in the properties of the powders and the size and complexity of the molds. Another example occurs in the assembly of printed circuit boards when there are either many or few common components to be placed on the circuit board. Different components need to be mounted on the insertion machines and this requires time.

The end result is that we now have a one-machine problem where the maximum completion time is no longer constant, but depends on the sequence of jobs. We ignore the processing times as they are constant and do not impact the choice of sequence. The inputs to such a problem are the setup times for each job as it may follow each other job. However, one must start with a job that does not have a predecessor. So we either have to assume a repeating cyclical schedule or be given a setup time for each job if it is the first job to be processed. It is much more common to assume a cyclical schedule and that is the case we will treat first. One reason for this is that the problem is identical to a very old and well-known problem called the traveling salesperson problem (TSP—minimal distance route of person travelling among a number of cities while only visiting each one once). One dissimilarity between the TSP and the sequence dependent setup (SDSU) problem is that the latter does not necessarily have a symmetrical matrix, that is, the setup of j after i may be different from the set up for i after j. In the TSP the distance between two cities is the same in either direction. These problems do not have a polynomial algorithm, but are amenable to solutions with dynamic programming or branch and bound. But first we will explore it with a simple greedy heuristic. Figure 14.1 contains the data for a four-job example. Always carefully note which are the following and leading jobs, as texts and problems may be presented differently in different sources.

The heuristic consist of selecting the smallest setup (arbitrary for ties) first, then selecting the next lowest feasible setup until we have one setup selected in each

Follower \ Predecessor	1	2	3	4
1	-	5	10	50
2	15	-	60	5
3	30	40	-	10
4	50	45	35	-

Figure 14.1 Sequence Dependent Setup (SDSU) Problem

Follower \ Predecessor	1	2	3	4
1	-	5	10	50
2	15	-	60	5
3	30	40	-	10
4	50	45	35	-

Figure 14.2 Heuristic Solution to the SDSU Example

row and each column. This is necessary so we do not have a job following two different jobs in the same sequence, an obvious impossibility. We first choose 2 following 4 (setup = 5). This eliminates the last column and the second row. Next we choose 1 following 2 (5 + 5 = 10). We are left with rows 3 and 4 and columns 1 and 3, from which we pick 3 following 1 (10 + 30 = 40). This leaves 4 following 3 (total setup = 40 + 35 = 75). Figure 14.2 shows the selected setups and you can verify that there is only one in each row and each column.

Next we apply dynamic programming. Because the solution is assumed to be cyclical, we are free to start with any one job. We will use the notation f(i,j) for the setup of i following j, where j may be an individual job or a group of jobs. As in our previous use of dynamic programming, we start with sets of one, then sets of two, until we arrive at the sets of the number of jobs in the problem. Arbitrarily choosing job 1 to start, the set of one are: f(2,1) = 15; f(3,1) = 30; f(4,1) = 50.

Sets of Two

f(2,3) = 60 + 30 = 90, that is if 2 follows 3 then 3 must follow 1 f(4,2) = 45 + 15 = 60
f(3,2) = 40 + 15 = 55; f(3,4) = 10 + 50 = 60; f(2,4) = 5 + 50 = 55; f(4,3) = 35 + 30

Sets of Three

f(2,(3,4) = min(60 + 60, 5 + 65) = 70,
that is, 2 follows 3, then 3 follows 4 or 2 follows 4, then 4 follows 3

f(3,(2,4) = min(40 + 55, 10 + 60) = 70
f(4,(2,3) = min(45 + 90, 35 + 55) = 90

And finally our one set of four, where one may be followed by each of the other three, and we look back to the sets of 3 to find the appropriate sequence.

f(1,(2,3,4) = min(5 + 70, 10 + 70, 50 + 90) = 75, which is our minimum

We trace the sequence back through the sets. The minimum for the set of four occurred when 1 followed 2. In that case 2 had to follow (3,4), which had given us 2 following 4, from which we see that 3 has to follow 4. Our final sequence is 3421. Because of the cyclical nature, this is equivalent to 4213, 2134 and 1342. If you had used complete enumeration you would have seen that each of these has a total setup of 75. In this case our simple heuristic happened to find the optimum.

We next proceed to the branch and bound approach. It is based on a few of the properties of our setup matrix. The first of these is the fact if we subtract the smallest number from each row and column, we remove amounts of setup from the problem that have to be part of the final sequence, but do not impact the choice of the optimum, which remains undetermined. The amount subtracted thus represents a lower bound. The second is that once we select a setup to enter our sequence, the symmetrical setup cannot be in the solution, as we stated earlier in the chapter. At any point in our procedure a zero element in the matrix is a potential choice to enter the solution. We begin by making the indicated subtractions from our original matrix, where the result is displayed in Figure 14.3. The amounts subtracted from the rows and columns are displayed on the margins of the matrix.

We next select a cell with a zero as our first potential entry in the sequence. We select the zero that has the highest sum of the minimum cells in its row and column. For example, the zero in (1,2) has a sum of 15, based on the cells (1,3) and (4,2). This is also the highest number among all the zeros, so we select it. By selecting it we eliminate every cell in its row and column form consideration, as well as its

Predecessor

Follower		1	2	3	4	
	1	-	0	5	45	5
	2	0	-	55	0	5
	3	10	30	-	0	10
	4	5	10	0	-	35
		10				65

Figure 14.3 Matrix after Subtracting from Each Row and Column. The Lower Bound is in the Lower Right Hand Corner

symmetrical partner (2,1). This will be our first branch. The second branch will be the use of its symmetrical partner. But let's continue with our first branch for now. See Figure 14.4. The zeros in cells (2,4) and (4,3) both have a 55 to subtract. Let's arbitrarily choose (4,3), thus adding (4,3) to our sequence, eliminating cells (2,3) and (4,1) as shown in Figure 14.5a. In Figure 14.5b, the last remaining subtraction has given us a lower bound of 75. We finally select cells (3,1) and (2,4) to complete our sequence. These had to be chosen because had we chosen (3,4) it would have eliminated all remaining zeros. The final sequence is shown in Figure 14.6. This is our trial solution of 1–3–4–2–1. We of course know that this is optimal from our dynamic programming. Had we not known, we would explore the other branches, starting with (2,1) until all branches had been fathomed.

Predecessor

Follower	1	2	3	4
1	-	*	-	-
2	-	-	55	0
3	10	-	-	0
4	5	-	0	-

65

Figure 14.4 The Matrix after Choosing (1,2)

a) Predecessor

Follower	1	2	3	4
1	-	*	-	-
2	-	-	-	0
3	10	-	-	0
4	-	-	*	-

65

b) Predecessor

Follower	1	2	3	4
1	-	*	-	-
2	-	-	-	0
3	0	-	-	0
4	-	-	*	-

10 75

Figure 14.5 SDSU Example after Subtraction in Column 1

Predecessor

Follower	1	2	3	4
1	-	*	-	-
2	-	-	-	*
3	*	-	-	-
4	-	-	*	-

Figure 14.6 Solution of the Non-Cyclical SDSU Problem

One other case of interest remains—when the solution is not cyclical and we have the initial setups given in Figure 14.7a. We modify our heuristic slightly by still selecting from four columns, but insisting that the initial column must be included. The result of this heuristic is shown in Figure 14.7b. This sequence is 2134 with a total setup of 95. When we apply dynamic program to this problem, we do get 3 optima, each with a total setup of 90, just slightly better than our heuristic. The calculations and the path back through the calculations to obtain one of the optima (2431) are shown in Figure 14.8 with the path outlined.

a)

Follower \ Predecessor	0	1	2	3	4
1	45	-	5	10	50
2	25	15	-	60	5
3	65	30	40	-	10
4	55	50	45	35	-

b)

Follower \ Predecessor	0	1	2	3	4
1	45	-	5	10	50
2	25	15	-	60	5
3	65	30	40	-	10
4	55	50	45	35	-

Figure 14.7 Non-Cyclical SDSU Problem and Heuristic Solution

f(1,0) = 45
f(2,0) = 25
f(3,0) = 65
f(4,0) = 55
f(1,2) = 5 + 25 = 30
f(2,1) = 15 + 45 = 60
f(1,3) = 10 + 65 = 75
f(3,1) = 30 + 45 = 75
f(1,4) = 50 + 55 = 105
f(4,1) = 50 + 45 = 95
f(2,3) = 60 + 65 = 125
f(3,2) = 40 + 25 = 65
f(2,4) = 5 + 55 = 60
f(4,2) = 45 + 25 = 70
f(3,4) = 10 + 55 = 65
f(4,3) = 35 + 65 = 100
f(1,(2,3)) = min(5 + 125 = 130 , 10 + 65 = 75) = 75
f(1,(2,4)) = min(5 + 60 = 65 , 50 + 70 = 120) = 65
f(1,(3,4)) = min(10 + 65 = 75 , 50 + 100 = 150) = 75
f(2,(1,3)) = min(15 + 75 = 90 , 60 + 75 = 135) = 90
f(2,(1,4)) = min(15 + 105 = 120 , 5 + 95 = 100) = 100
f(2,(3,4)) = min(60 + 65 = 125 , 5 + 100 = 105) = 105
f(3,(1,2)) = min(30 + 30 = 60 , 40 + 60 = 100) = 60
f(3,(1,4)) = min(30 + 105 = 135 , 10 + 95 = 105) = 105
f(3,(2,4)) = min(40 + 60 = 100 , 10 + 70 = 80) = 80
f(4,(1,2)) = min(50 + 30 = 80 , 45 + 60 = 105) = 80
f(4,(1,3)) = min(50 + 75 = 125 , 35 + 75 = 110) = 110
f(4,(2,3)) = min(45 + 125 = 170 , 35 + 65 = 100) = 100
f(1,(2,3,4))= min(5 + 105 = 110 , 10 + 80 = 90 , 50 + 100 = 150) = 90
f(2,(1,3,4))= min(15 + 75 = 90 , 60 + 105 = 165 , 5 + 110 = 115) = 90
f(3,(1,2,4))= min(30 + 65 = 95 , 40 + 100 = 140 , 10 + 80 = 90) = 90
f(4,(1,2,3))= min(50 + 75 = 125 , 45 + 90 = 135 , 35 + 60 = 95) = 95

Figure 14.8 Dynamic Programming Solution to the Non-Cyclical SDSU Problem

Early/Tardy Problem

Another class of one-machine problems that is of interest to us is the collection of problems that considers penalties for both earliness as well as lateness. These are referred to as early/tardy problems. These surfaced during the last 30 years with the interest in the just in time (JIT) philosophy of operations (Baker and Scudder, 1990). One aspect of this approach is that it is wasteful to complete jobs before they are required, because inventory then has to be held from completion to delivery. We assume that a client is neither willing to accept early delivery nor willing to pay for it before the due date.

We treat these problems separately from other one-machine problems because the measure is not regular—when an early job is rescheduled to be less early, its completion time increases while the measure decreases. This violates the definition of regularity.

One common characteristic of these problems is that as the earliness of jobs becomes larger, the penalty is increased and as the tardiness increases so does the penalty for it. This gives rise to a V-shaped penalty function as shown in Figure 14.9.

We begin with the most basic example, equal or unequal penalties for earliness and tardiness and a common due date for a one-machine problem. Common due dates are not that rare—think of loading an airplane, getting all the parts ready for an assembly, loading a truck—these are all examples of common due dates. Because the earlier a job finishes, the larger its early penalty becomes, it is best to keep the processing times close to the due date small, thus creating an SPT sequence going back in time from the due date. Similarly, we prefer to have short processing times after the due date, creating an SPT schedule of jobs going forward from the due date. This scheme does indeed create optimal schedules as long as we can place enough jobs between the due date and time 0 and one job finishes at the due date (Baker, 2009). If this is possible we have what is called an unrestricted problem. But how do we decide which jobs are to be processed before the due date? We will need some definitions. The notation $|B|$ indicates the size of the set B. Let the set B be the jobs placed before the due date and A the set of jobs placed after the due date. We also define the penalty for earliness as α and the penalty for tardiness as β. We wish to minimize:

$$\sum_{i=1}^{n}\left(\alpha\left(max\left(0,d_i-C_i\right)\right)+\beta\left(max\left(0,-d_i+C_i\right)\right)\right) \qquad \text{Equ. 14.1}$$

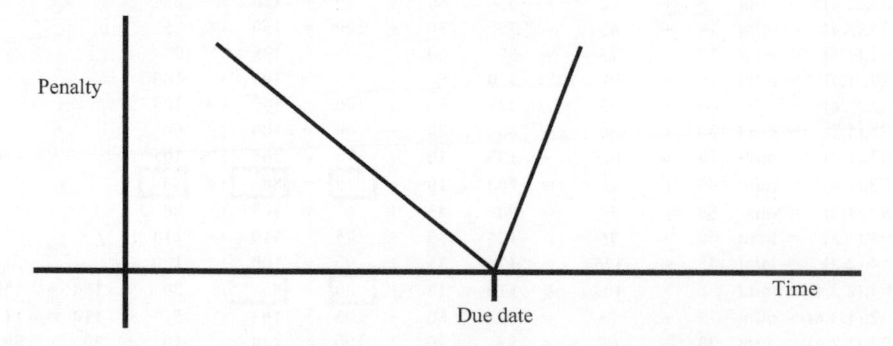

Figure 14.9 V-Shaped Penalty—Unequal Penalties for Earliness and Tardiness

Naturally, before we start placing jobs, both sets are empty and |A| = |B| = 0. The following algorithm places the jobs.

Algorithm 14.1

Step 1: Order all jobs in non-increasing order (LPT).

Step 2: If $\alpha|B| < \beta(1+|A|)$ then place jobs from the beginning of the list in set B, else in set A.

Step 3: Fill the final sequence from the list created in step 2 with jobs from set B at the beginning of the sequence and with jobs from set A from the end of the sequence. Fix the due date at the completion of the last job in B.

Step 4: The schedule starts at the beginning of the first job in set B and is optimal if the time between the due date and the current time is as at least as long as the sum of the processing times in set B. If this is not the case, the problem is called restricted and heuristic methods are needed to create a good schedule.

Example 14.1 Given 22 jobs with the penalty for earliness = 1 and the penalty for tardiness = 3, determine an optimal sequence where the common due date is 145. The initial jobs and the solution are shown in Figure 14.10.

One may determine the number of jobs before the due date by

$$|B| = \text{smallest integer} > \frac{n\beta}{\alpha+\beta} \qquad \text{Equ. 14.2}$$

Job	p	Jobs in LPT	p	B	A	alphaB	beta(1+A)	Assign to	Job Sequence	p	Completion	Penalty
											0	
1	8	21	18	21		0	3	B	21	18	18	127
2	7	14	15	14		1	3	B	14	15	33	112
3	5	22	13	22		2	3	B	22	13	46	99
4	9	11	12		11	3	3	A	18	12	58	87
5	3	18	12	18		3	6	B	20	11	69	76
6	4	20	11	20		4	6	B	4	9	78	67
7	3	4	9	4		5	6	B	1	8	86	59
8	8	19	9		19	6	6	A	8	8	94	51
9	4	1	8	1		6	9	B	13	8	102	43
10	7	8	8	8		7	9	B	2	7	109	36
11	12	13	8	13		8	9	B	10	7	116	29
12	7	17	8		17	9	9	A	12	7	123	22
13	8	2	7	2		9	12	B	13	8	131	14
14	15	10	7	10		10	12	B	6	4	135	10
15	5	12	7	12		11	12	B	9	4	139	6
16	3	3	5		3	12	12	A	7	3	142	3
17	8	15	5	15		12	15	B	16	3	145	0
18	12	6	4	6		13	15	B	5	3	148	9
19	9	9	4	9		14	15	B	3	5	153	24
20	11	5	3		5	15	15	A	17	8	161	48
21	18	7	3	7		15	18	B	19	9	170	75
22	13	16	3	16		16	18	B	11	12	182	111
											Total Penalty	1108

Figure 14.10 Early/Tardy Problem with Linear Penalties

If the problem is restricted, the simplest approach is to delay the start and let more jobs be tardy. Another one is to make the due date a decision variable and allow it to be later. Many other objective functions are possible (Baker, 2009), but are beyond an introduction to this topic.

Batch Sequencing

This section treats a very common problem that occurs when many products are processed on the same production line. Therefore this is a one-machine problem, even though the line may consist of many machines. But because the machines are essentially tied together, we can treat this as a one-machine problem. Frequently groups of jobs use the same setup, so there is a substantial advantage to process these jobs as a single group. We assume that each job has its own ready time, due date, and processing time, while a group shares a setup time.

Our objective is to minimize the average (or total) tardiness. Unfortunately there is no known polynomial algorithm for this kind of problem. We do have a heuristic. The basic idea is to keep adding jobs to the same group as long as we do not cause a job in another group to be tardy. We begin by sorting the jobs in non-decreasing order of ready times and due dates. In case of ties, one should probably choose the shorter processing time and consequently reduce the average flow time, although this is not our primary objective. As an aid in determining whether to add another job to a batch, we may calculate the latest that a job may start after having to be set up and not be tardy:

$$\text{Latest start time}\left(\text{LST}\right) = d_i - \text{setup} - p_i \qquad \text{Equ. 14.3}$$

The heuristic:

1. Order all jobs by group, non-increasing ready time, due date and processing time, starting with the group that has the job with the earliest ready time, followed by the group with next lowest ready times, and so forth.
2. Start creating a sequence with the first job and keep adding jobs from the same group.
3. As you add jobs, check to see if any jobs from the other groups would be tardy as you add that job (calculate the LST).
4. If a job is made tardy, do not add the job, but start the second group.
5. Repeat this process until all jobs are processed. Calculate the total tardiness.
6. Look for exchanges that may reduce the total tardiness.

As you can surmise, it will generally not be possible to keep all the jobs in a group together in a sequence nor to not exceed the LST. Although the maximum completion time is not our primary objective, we can assess a schedule by calculating it and compare it to both an upper and lower bound. The lower bound is the sum of one setup time for each group plus the total processing time of all jobs. The upper bound is the sum of the setups for each job plus the total processing time.

Because the sum of all processing times is a constant, we really only have to consider the setups.

If we are also interested in allowing the maximum completion to increase in order to reduce the total tardiness, we can experiment with adjusting the sequence. Applying a Monte Carlo simulation (Chapter 12) with random exchanges will usually yield good results.

Example 14.2 Consider 14 jobs that belong to four different groups as shown in Figure 14.11. Application of the heuristic gives a C_{max} of 75 and a total tardiness of 128 (see Figure 14.12).

Job	Type	Processing time	Arrival time	Due date	start	finish	tardy	Early	LST
					0				
1	A	3	0	17	3	6	0	11	11
2	A	6	0	9	6	12	3	0	0
3	C	6	0	19	16	22	3	0	9
4	C	3	0	20	22	25	5	0	13
5	D	2	18	30	28	30	0	0	25
6	D	4	18	43	30	34	0	9	36
7	D	4	32	65	34	38	0	27	58
8	A	2	4	15	41	43	28	0	10
9	B	5	55	72	55	60	0	12	65
10	B	6	40	55	60	66	11	0	47
11	A	5	5	25	69	74	49	0	17
12	C	6	4	20	78	84	64	0	10
13	B	5	22	75	86	91	16	0	68
14	B	4	5	18	91	95	77	0	12
	Total						256		

Setup times	
A	3
B	2
C	4
D	3

Figure 14.11 Batch Sequencing Example

	2	3	4	5	start	finish	tardy	Early	LST
Job	Type	Processing time	Arrival time	Due date	0				
2	A	6	0	9	3	9	0	0	0
1	A	3	0	17	9	12	0	5	11
8	A	2	4	15	12	14	0	1	10
11	A	5	5	25	14	19	0	6	17
3	C	6	0	19	23	29	10	0	9
4	C	3	0	20	29	32	12	0	13
12	C	6	4	20	32	38	18	0	10
14	B	4	5	18	40	44	26	0	12
13	B	5	22	75	44	49	0	26	68
10	B	6	40	55	49	55	0	0	47
7	D	4	32	65	58	62	0	3	58
5	D	2	18	30	62	64	34	0	25
6	D	4	18	43	64	68	25	0	36
9	B	5	55	72	70	75	3	0	65
125	Total						128		

Figure 14.12 Batch Sequencing Example after Applying the Heuristic

Batch Processing

While batch processing sounds very similar to batch sequencing, they are quite different in the kind of problem they apply to and their method of solution. Batch processing is frequently referred to as the baking problem. While there are many similarities to baking bread and pizzas, there are also differences. These problems often arise in the manufacture of integrated circuits, LCD panels, and chemical processing, as well as metallurgy. This is another instance of a one-machine problem, the oven. The basic parameters of this problem are:

1. An oven, or a similar device that has a capacity to hold a specific number of jobs of a given type. The capacity for different types may be different.
2. Each type of job requires a specific set of conditions in the oven. Pressure, temperature, and time are good examples.
3. The processing in the oven is such that it may not be interrupted before completion. This prevents us from adding additional jobs to a partially filled oven once it has started.
4. Only jobs of the same type may share the oven. That means that jobs in the oven require the same pressure, temperature, and time.

Our objective is the same as in batch sequencing—minimize total tardiness, with a secondary objective of minimizing maximum completion time and average flow time. We have no polynomial time algorithm and resort to heuristics that are not very well defined, but nevertheless can improve on some initial solutions. The processing of a group of jobs in the oven is considered a batch.

Heuristic

1. Once again we order the jobs in non-increasing ready times and due dates.
2. From this sort we then begin to load a batch with the same type of jobs until the oven is full and then start the next batch. Each batch's start time is the completion of the previous batch.
3. If all batches are full except perhaps the last one and no jobs are tardy we have reached an optimum, which is unlikely in most cases.
4. We continue by perhaps letting some batches start before they are full when the wait for the ready time of the next job of the same type is somewhat later.

Example 14.3 Given 20 jobs of three types and an oven capable of holding four jobs, sequence the jobs to minimize total tardiness (Figure 14.13).

The example after the application of the heuristic is shown in Figure 14.14.

	Processing time	Oven capacity
Job type A	1.5	4
Job type B	2	4
Job type C	2.5	4

Job	Type	Arrival	Due date
1	A	0	5
2	A	0	8
3	B	6	12
4	A	1	8
5	B	4	17
6	B	5	13
7	C	2	7
8	A	5	12
9	C	8	13
10	B	10	16
11	B	11	16
12	C	2	8
13	C	9	13
14	A	3	8
15	A	7	14
16	B	9	15
17	C	1	5
18	B	6	15
19	A	8	15
20	A	9	14

Figure 14.13 Batch Processing Example

Schedule2	Batch	Job	Arrival	Process time	Loading	Completion	Tardy	Early
A	1	1	0	1.5	1	1.5	0	3.5
A	1	2	0	1.5	2	1.5	0	6.5
A	1	4	1	1.5	3	1.5	0	6.5
A	1	14	3	1.5	4	1.5	0	6.5
C	2	17	1	2.5	1	4	0	1
C	2	7	2	2.5	2	4	0	3
C	2	12	2	2.5	3	4	0	4
C	2	9	8	2.5	4	4	0	9
B	3	5	4	2	1	6	0	11
B	3	6	5	2	2	6	0	7
B	3	3	6	2	3	6	0	6
B	3	18	6	2	4	6	0	9
A	4	8	5	1.5	1	7.5	0	4.5
A	4	15	7	1.5	2	7.5	0	6.5
A	4	19	8	1.5	3	7.5	0	7.5
A	4	20	9	1.5	4	7.5	0	6.5
B	5	16	9	2	1	9.5	0	5.5
B	5	10	10	2	2	9.5	0	6.5
B	5	11	11	2	3	9.5	0	6.5
C	6	13	9	2.5	1	12	0	1

Figure 14.14 Heuristic Applied to the Batch Processing Example

Net Present Value

Although not a relaxation of an assumption, this is a good place to include this rarely mentioned approach to scheduling. It is based on the time concept of the value of money. We can apply it to a schedule because different events in a schedule represent either income or expense or both at different points in time. Figure 14.15 details one set of assumptions we can make about when we either pay for something or receive an income. For the purposes of this analysis we ignore labor costs. A specific schedule will then represent a series of payments or receipts and we can

Event		Action
Ready time		Pay for material
Completion time		
	On time	Receive payment from customer
	Before due date	Pay for carrying finished goods inventory until due date
	After due date	Receive payment from customer
		Pay late fee
Flow time		Pay for carrying work in process inventory

Figure 14.15 Net Present Value Definitions

Job	1	2	3	4	5	6	7
p	3	2	4	5	3	2	3
d	6	12	19	7	11	26	23
Material Cost	$20	$30	$25	$15	$35	$11	$18
Finished Good value	$100	$150	$122	$88	$130	$143	$120
Late fee/period	$0.50	$0.75	$0.60	$0.40	$0.65	$0.80	$0.65
Annual Rate of return	20%						
Inventory carrying rate	25%						

Figure 14.16 Example 14.4

Job		1	4	5	2	3	7	6	
p		3	5	3	2	4	3	2	
d		6	7	11	12	19	23	26	
Ready		0	1	5	6	13	17	20	
Start	0	0	3	8	11	13	17	20	
Completion		3	8	11	13	17	20	22	
Tardy		0	1	0	1	0	0	0	
Early		3	0	0	0	2	3	4	
Flow		3	7	6	7	4	3	2	
									Totals
		0	0.4	0	0.75	0	0	0	
NPV Tardy		$ -	$ 0.388	$ -	$ 0.713	$ -	$ -	$ -	$ 1.101
Early cost		$ 0.0144	$ -	$ -	$ -	$ 0.0096	$ 0.0144	$ 0.0192	
NPV Early		$ 0.0143	$ -	$ -	$ -	$ 0.0090	$ 0.0134	$ 0.0177	$ 0.054
Material		$ 20	$ 15	$ 35	$ 30	$ 25	$ 18	$ 11	
NPV material Cost		$ 20.00	$ 14.94	$ 34.33	$ 29.32	$ 23.78	$ 16.86	$ 10.19	$ 149.427
Receipt		$ 100.00	$ 88.00	$ 130.00	$ 150.00	$ 122.00	$ 120.00	$ 143.00	
NPV receipt		$ 97.72	$ 85.34	$ 124.62	$ 142.70	$ 113.42	$ 109.86	$ 129.42	$ 803.079
lead time	6							Net	$ 652.50

Figure 14.17 Solution for Example 14.4

calculate a present value of these transaction, assuming a desirable rate of return (see any text on engineering economy, such as Park (2015)). If we change the schedule, the net present value will also change. We can then use a heuristic to find a good schedule by attempting to maximize the net present value. From a practical point of view, we should approach it from an MRP perspective; that is, use due dates and a constant lead time to determine the ready times. The constant lead time then becomes another variable to be determined. The role of the ready times is important as it not only determines the eventual inventory held during production, but also our ability to meet the due dates. A long lead time assumption gives rise to higher inventory, but at the same time reduces tardiness.

Example 14.4 Given the data in Figure 14.16, determine a good schedule for this seven-job, one-machine problem, using a preferred annual rate of return of 20%. The periods are in weeks. The solution obtained using Excel's Solver is shown Figure 14.17.

15 DYNAMIC AND STOCHASTIC PROBLEMS

Introduction

In Chapter 6 we defined dynamic and stochastic problems as follows: Problems in which jobs arrive randomly over a period of time are called *dynamic*. Problems in which the processing times, etc. are uncertain are called *stochastic*. In this chapter we look at a few cases for which there are algorithms available. The vast majority of such problems need to be treated with simulation, of which we will give some examples. We begin with a brief review of the required statistics.

Recall that a density function f(t) and the corresponding distribution are defined by:

$$F(t) = P(X < t) = \int_0^t f(t)dt \qquad \text{Equ. 15.1}$$

Among others, we will be using the exponential distribution, where $f(t) = e^{-\lambda t}$, and whose graphs for $\lambda = 0.2$ are shown in Figure 15.1. We will also use the normal and uniform distributions. As in previous problems, we will not consider preemption. One major difference from static and deterministic problems is that we will be considering random values of all our stochastic variables, and as a consequence will be dealing with expected values and standard deviations of our measures. Most commonly, we use the exponential distribution for the processing times because we have algorithms for it. For other distributions we have to use simulations. The expected value for the exponential distribution is:

$$E(X) = \int_0^\infty tf(t)dt = \frac{1}{\lambda} \qquad \text{Equ. 15.2}$$

Suppose we have two jobs to process on a single machine. Job 1 has an expected processing time of 2 and job 2 has an expected processing time of 3. These are distributed exponentially. The expected completion time of both jobs is 2 + 3 = 5. A simple simulation of the completion time is shown Figure 15.2. Note that while the individual values vary extensively, the averages are fairly close to the expected values even for such a small sample.

There are a number of theorems related to dynamic and stochastic problems. We distinguish between list and dynamic policies. A list policy assumes that we have a list of jobs that we are to sequence. A dynamic policy assumes that jobs arrive randomly.

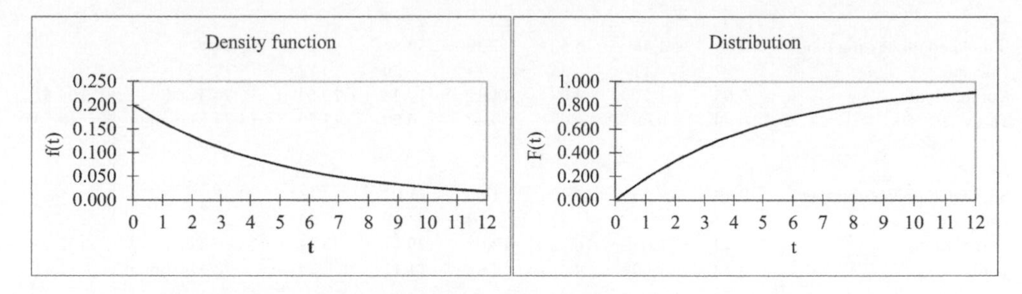

Figure 15.1 The Exponential Distribution

Trial	Job1	Job2	Sum
1	4.19	5.22	9.41
2	0.61	1.01	1.62
3	1.04	1.18	2.22
4	0.02	5.83	5.85
5	1.10	2.51	3.61
6	6.63	0.28	6.92
7	1.13	2.09	3.22
Average	2.10	2.59	4.69
Std. dev.	2.40	2.14	2.81

Figure 15.2 Simulated Values for the Two-Job, One-Machine Problem

Job	1	2	3	4	5	6	7		
Expected processing time	2	5	2	6	3	3	7		
Due date	7	29	11	12	26	14	18		
Job	1	3	5	6	2	4	7		
Expected processing time	2	2	3	3	5	6	7		
Due date	7	11	26	14	29	12	18		
Completion	7	9	12	15	20	26	33	Total	122
Tardy	0	0	0	1	0	14	15	Max	15
Job	1	3	4	6	7	5	2		
Expected processing time	2	2	6	3	7	3	5		
Due date	7	11	12	14	18	26	29		
Completion	7	9	15	18	25	28	33	Total	135
Tardy	0	0	3	4	7	2	4	Max	7

Figure 15.3 Static SPT and T_{max} Example

Theorem 15.1 For the static and deterministic single machine the SPT rule minimizes $\overline{F}$. Similarly, for both list and dynamic policies, for arbitrary distributions of processing times or arrival times, a rule of shortest expected processing times (SEPT) will minimize the expected $\overline{F}$.

Theorem 15.2 For the static and deterministic single machine the EDD rule minimizes L_{max}. For both list and dynamic policies, for arbitrary distributions of processing times or arrival times and deterministic due dates, EDD will minimize the expected value of L_{max}.

Figure 15.3 has an example of a one-machine problem with stochastic processing times. It also has the solution for $\sum F$ and T_{max} using static processing times.

Job	1	3	5	6	2	4	7		
Simulated processing time	0.05	1.46	5.52	7.39	4.82	6.27	11.24		
Due date	7	11	26	14	29	12	18		
Completion	0.05	1.51	7.03	14.42	19.24	25.50	36.74	Total	104.51
Tardy	0.00	0.00	0.00	0.42	0.00	13.50	18.74	Maximum	18.74

Job	1	3	4	6	7	5	2		
Simulated processing time	9.49	2.50	4.23	1.86	11.03	4.13	0.66		
Due date	7	11	12	14	18	26	29		
Completion	9.49	12.00	16.22	18.08	29.11	33.24	33.90	Total	152.04
Tardy	2.49	1.00	4.22	4.08	11.11	7.24	4.90	Maximum	11.11

Figure 15.4 Stochastic SPT and T_{max} Example

	SPT		EDD	
	Sum of F	Tmax	Sum of F	Tmax
Average	89.64	13.07	95.28	5.29
Std. Dev.	32.85	9.61	41.88	7.44

Figure 15.5 100 Simulations Results for the Stochastic SPT and T_{max} Example

Job	1	2	3	4	5	6	7	8	9
Expected Processing time	8	4	1	5	6	7	3	2	10

	Job	Completion	Job	Completion	Job	Completion	Job	Completion	Job	Completion
Machine 1	9	10	5	16	2	20	7	23		
Machine 2	1	8	6	15	4	20	8	22	3	23

	Job	Completion	Job	Completion	Job	Completion	Job	Completion	Job	Completion
Machine 1	9	30.5	5	32.3	2	33.6	7	35.2		
Machine 2	1	6.7	6	8.7	4	12.5	8	17.2	3	20.4

Figure 15.6 Parallel Example

Figure 15.4 has the same sorting by expected processing and due date, but with an instance of a simulation of the processing times using the exponential distribution {using LN(1 – rand())*expected processing time) in Excel}. The simulation has also been repeated 100 times and the averages for $\sum F$ and T_{max} and are displayed in Figure 15.5. Because we are using exponential distribution, the standard deviations approach the averages.

Theorem 15.3 For the static and deterministic parallel machines the LPT rule tends to minimize C_{max}. For the list policy, for exponential distributions of processing times, a rule of longest expected processing time will tend to minimize the expected C_{max} for 2 machines.

It is important to note that here we only say that it will tend, not actually achieve the optimum, just as we saw in Chapter 13. Figure 15.6 gives an example of a nine-job, two-machine parallel problem, with static and stochastic solutions. The average of 100 simulation was 23.9, with standard deviation of 11.2. Note that the averages of the static and stochastic cases are very close.

Before we look at stochastic flow shops, we need to define blocking. By blocking we mean that a completed operation on the first machine cannot leave the machine because machine 2 is not finished processing its operation. This prevents a flow line with equal but stochastic processing times on each machine, from realizing a production rate equal to the reciprocal of the processing time. Consider two machines with an average processing time of 3 each and a standard deviation of 0.2 with a normal distribution that is truncated at 3 standard deviations. If the two were independent of each other, then the completion on both machines would be as one can observe on the left of Figure 15.7, while the blocking case, without a buffer between the machines, is on the right hand side. Both cases used the same random values for each of the machines. The blocking case has longer total processing time in all of 100 simulations, indicating that buffers are necessary when processing times vary. The averages and standard deviations of the simulations are shown in Figure 15.8 that show a consistent effect of blocking.

This leaves us with one more important question for this chapter. How should we sequence a number of jobs on two machines with stochastic processing to minimize the expected maximum completion time? Let the processing time of job i on machine 1 be X_{i1} (a random variable replacing p_{i1}) with an exponential distribution with rate λ_i. Let the processing time of job i on machine 2 be X_{i2} (a random variable replacing p_{i2}) with an exponential distribution with rate μ_i.

	Assuming an adequate buffer			
Sample	First Machine	Second Machine	Completion	Total Time
1	2.723	3.483	6.206	
2	2.782	3.234	6.016	9.44
3	2.98	3.132	6.112	12.572
4	2.938	3.074	6.012	15.646
5	3.277	3.007	6.284	18.653
6	2.993	2.814	5.807	21.467
7	3.127	3.141	6.268	24.608
8	3.166	3.097	6.263	27.705
9	3.083	3.271	6.354	30.976
10	3.06	3.06	6.12	34.036
11	3.1	3.466	6.566	37.502
12	2.876	3.066	5.942	40.568
13	2.913	3.187	6.1	43.755
14	3.132	3.058	6.19	46.813
15	3.313	2.697	6.01	49.51
16	3.128	2.771	5.899	52.281
17	2.962	3.065	6.027	55.346
18	3.05	2.89	5.94	58.236
19	3.153	2.866	6.019	61.102
20	2.849	3.357	6.206	64.459
Average	3.030	3.087	6.117	
Std. Dev.	0.155	0.214	0.179	

	Accounting for Blocking					
Sample	Start	First Machine	Finish	Start	Second Machine	Finish
1	0	2.723	2.723	2.723	3.483	6.206
2	2.723	2.782	6.206	6.206	3.234	9.44
3	6.206	2.98	9.44	9.44	3.132	12.572
4	9.44	2.938	12.572	12.572	3.074	15.646
5	12.572	3.277	15.849	15.849	3.007	18.856
6	15.849	2.993	18.856	18.856	2.814	21.67
7	18.856	3.127	21.983	21.983	3.141	25.124
8	21.983	3.166	25.149	25.149	3.097	28.246
9	25.149	3.083	28.246	28.246	3.271	31.517
10	28.246	3.06	31.517	31.517	3.06	34.577
11	31.517	3.1	34.617	34.617	3.466	38.083
12	34.617	2.876	38.083	38.083	3.066	41.149
13	38.083	2.913	41.149	41.149	3.187	44.336
14	41.149	3.132	44.336	44.336	3.058	47.394
15	44.336	3.313	47.649	47.649	2.697	50.346
16	47.649	3.128	50.777	50.777	2.771	53.548
17	50.777	2.962	53.739	53.739	3.065	56.804
18	53.739	3.05	56.804	56.804	2.89	59.694
19	56.804	3.153	59.957	59.957	2.866	62.823
20	59.957	2.849	62.823	62.823	3.357	66.18
		3.030			3.087	
		0.155			0.214	

Figure 15.7 Flow Example without and with Blocking

	Buffered	Blocking
Average	63.12	65.22
Stdev	0.92	0.69

Figure 15.8 Result of 100 Simulation of Blocking vs. Buffering

Sorted by Theorem 15-4

				Exponential Sample		Stochastic		Expected Processing		Deterministic	
Job	λ	μ	λ-μ	X1	X2	C1	C2	P1	P2	C1	C2
8	1.00	0.75	0.25	0.49	0.79	0.49	1.28	1.00	1.33	1.00	2.33
6	0.82	0.67	0.15	4.01	1.05	4.50	5.54	1.22	1.49	2.22	3.83
7	0.21	0.08	0.13	3.17	25.58	7.67	33.25	4.76	12.50	6.98	19.48
3	0.90	0.82	0.08	1.19	1.20	8.86	34.45	1.11	1.22	8.09	20.70
9	0.57	0.49	0.08	4.58	0.70	13.44	35.15	1.75	2.04	9.85	22.74
2	0.13	0.23	-0.10	2.20	0.87	15.63	36.02	7.69	4.35	17.54	27.09
5	0.33	0.48	-0.15	6.67	2.81	22.31	38.84	3.03	2.08	20.57	29.17
1	0.26	0.72	-0.46	1.70	0.69	24.00	39.52	3.85	1.39	24.42	30.56
4	0.42	0.93	-0.51	1.34	0.85	25.34	40.37	2.38	1.08	26.80	31.64

Sorted by Johnson's method

				Exponential Sample		Stochastic		Expected Processing		Deterministic	
Job	λ	μ	λ-μ	X1	X2	C1	C2	P1	P2	C1	C2
8	1.00	0.75	0.25	0.51	2.37	0.51	2.88	1.00	1.33	1.00	2.33
3	0.90	0.82	0.08	0.01	1.00	0.52	3.88	1.11	1.22	2.11	3.55
6	0.82	0.67	0.15	0.70	0.07	1.22	3.95	1.22	1.49	3.33	5.05
9	0.57	0.49	0.08	3.15	1.87	4.38	6.25	1.75	2.04	5.09	7.13
7	0.21	0.08	0.13	2.75	45.82	7.13	52.94	4.76	12.50	9.85	22.35
2	0.13	0.23	-0.10	21.69	9.07	28.82	62.01	7.69	4.35	17.54	26.69
5	0.33	0.48	-0.15	6.20	3.98	35.02	65.99	3.03	2.08	20.57	28.78
1	0.26	0.72	-0.46	13.70	2.47	48.72	68.46	3.85	1.39	24.42	30.17
4	0.42	0.93	-0.51	3.87	0.19	52.58	68.65	2.38	1.08	26.80	31.24

Figure 15.9 Flow Shop Performance

	Theorem	Johnson
Average	34.46	39.55
Std. Dev.	11.71	7.81

Figure 15.10 Result of 100 Simulation of Flow Shop

Smallest Variance

			Stochastic				Deterministic			
Job	Mean	Variance	X1	X2	C1	C2	P1	P2	C1	C2
3	2.00	0.23	2.28	2.35	2.28	4.63	2.00	2.00	2.00	4.00
7	2.00	0.28	2.33	2.22	4.62	6.85	2.00	2.00	4.00	6.00
8	2.00	0.30	1.91	1.74	6.52	8.59	2.00	2.00	6.00	8.00
1	2.00	0.40	2.36	1.96	8.88	10.84	2.00	2.00	8.00	10.00
5	2.00	0.62	1.46	1.58	10.34	12.42	2.00	2.00	10.00	12.00
6	2.00	0.87	3.39	0.04	13.72	13.76	2.00	2.00	12.00	14.00
2	2.00	0.89	1.59	0.75	15.31	16.06	2.00	2.00	14.00	16.00
4	2.00	0.98	2.68	2.24	17.99	20.22	2.00	2.00	16.00	18.00
9	2.00	0.99	2.25	3.51	20.24	23.75	2.00	2.00	18.00	20.00

Figure 15.11 Smallest Variance Example

Theorem 15.4 Sequencing the jobs in decreasing order of $\lambda_i - \mu_i$ will minimize the expected makespan. Figures 15.9 and 15.10 show the effect of using the theorem instead of using Johnson's theorem. The sequencing indicated by Theorem 15.4 performs only slightly worse for deterministic sequencing, while giving an improvement with stochastic sequencing.

If the processing times are equal, then of course Theorem 15.4 is of no use. However, if the variances are different, we can apply the smallest variance first rule, as in Theorem 15.5.

Theorem 15.5 The smallest variance first (SV) rule minimizes the expected completion time. An example, using normal distributions for the processing times for this theorem, is shown in Figure 15.11.

APPENDIX A

Costing of Products and Services

Planning and scheduling both have a substantial impact on the cost of our products and services. We start with distinguishing among predicted cost, actual cost, and price. Organizations need to assess their predicted cost in order to set a price that will allow them to make an adequate profit and be consistent with the price that the market will bear. The market will, unless they have a monopoly or completely differentiated product, dictate a rather narrow price range for their product. In the end the organization also needs to assess the actual cost incurred, not only to know which of the products or services were profitable, but also to verify their costing procedures. As we have mentioned, the language is that of manufacturing here also, but of course the methods are equally applicable to services. Even though different organizations have different ways of determining their costs, the method we present here is fairly typical. These methods of predicting cost are based on a number of definitions:

1. Direct material cost per product. This is the cost incurred when material or goods are purchased from a source other than the organization in question. These are well-defined, as they have to be paid for at prearranged prices.
2. Annual total purchased material. This is the sum of all material purchased for all products that the organization manufactures in a year.
3. Annual total material overhead. These are the costs associated with acquiring, storing, insuring, and moving material by people not directly involved in producing the product. These expenses have to be allocated to specific products to account for the total cost of a product.
4. Direct labor cost per hour. This refers to the cost of labor that is expended in producing the product and is limited to people who actually touch the product or provide the service. It usually includes benefits paid to these people. These data are also well-defined.
5. Amount of labor required to produce a product. This is a precise measure, determined by observation or software, of how long it takes a person or persons working continuously and efficiently to produce the product.
6. Total annual labor hours. This number is arrived at by multiplying the labor required by each product with the number of units to be produced in a year and summing over all products.

7. Labor productivity. Because people (and sometimes even machines) cannot work continuously, we define a percentage, most commonly 85%, by which we have to inflate the time to produce the product to account for breaks, random inefficiencies, unavailable materials, defective parts, machine breakdowns, fatigue, and so forth.
8. Total annual indirect overhead. This includes everything else not mentioned in items 1–7. Engineering, accounting, supervision, all facility costs, equipment depreciation, utilities, real estate taxes, and so forth are part of this category. It does not include the items mentioned under material overhead, although some organizations combine the two.

Example A.1 Consider a product that requires \$15 worth of material and 0.2345 hours of labor. The organization that produces this product purchases \$200 million for all their products in direct materials per year, while accumulating \$3 million in material overhead in the year. The relevant factors for direct labor are direct labor cost of \$30/hour, 280,000 hours of required labor for all their products, \$24 million in indirect overhead and a labor efficiency of 85%.

Solution We begin by calculating the material overhead rate—the additional cost incurred on every purchased item.

material overhead rate = material overhead/total material purchased
= \$3M/\$200M = 1.5% Equ. A.1

Therefore the material cost of this product is (1 + 1.5%)★\$15 = \$15.23 (some organizations will carry this number to more decimal places, depending on the number of units to be produced). Next we allocate the indirect overhead to the basic labor rate:

burdened (total or loaded) labor rate =
(indirect overhead + (total labor hours★direct labor rate)/ labor efficiency)/total labor hours
= (\$24M + (280,000★\$30)/(85%)/280,000
= \$121.01 Equ. A.2

While this may seem high to you, try to remember the last time you or a friend had a car repaired. The labor rate quoted was probably somewhere around \$100/hour, certainly substantially more than the mechanic's pay. It remains to calculate the cost of the product:

material cost + labor cost = \$15.23 + loaded labor rate★hours per product
= \$15.23 + \$121.01★0.2345
= \$15.23 + \$28.38
= \$43.60 Equ. A.3

Many other ways are possible to allocate costs, as you will see in the problems for this section (Appendix E).

		%
Material	$200,000,000	85%
Indirect overhead	$24,000,000	10%
Direct labor	$8,400,000	4%
Material Overhead	$3,000,000	1%
Total	$235,400,000	

Figure A.1 Distribution of Costs in Our Example Organization

Also, the relationship of various items will vary considerably across organizations. For example, an organization that manufactures most of its own parts will have lower material but higher labor costs. Figure A.1 shows the relationships for this particular organization.

APPENDIX B

Project Scheduling

One of the common applications of scheduling is in the determination of completion times of projects. Most of this text dealt with tasks that were repetitive. Projects, on the other hand, while they may have some similarities to each other, are generally unique and executed only once. Formal approaches to scheduling projects date from the 1950s and fall into two categories—those that are sufficiently new so that the duration of individual tasks is very variable (generally referred to as PERT—program evaluation and review technique), and those that have tasks that are quite predictable in duration (generally referred to as CPM—critical path method). However, the method of each may easily be applied to the other.

There are many commercial software packages available to analyze and manage all aspects of projects. Many books also treat projects in great detail. Our only purpose here is to familiarize the reader with the methods of determining the longest path through the network that represents a project and the resulting earliest possible completion of the project. In that sense it is quite similar to the shifting bottleneck method that was discussed in Chapter 12. The main difference, as we shall see shortly, is that we require two passes through the network.

We will need some definitions. There are two basic types of representation of the activities (also referred to as tasks or jobs) of a project.

The first of these is called activity on arc (AOA), where we represent the duration of a task with a directed arc from a node to another node in the network. There is a starting node from which all original tasks originate and a final node at which all final tasks terminate. We call every node an event. Every task has predecessors and successors. An activity cannot start until all of its predecessors have been completed. See Figure B.1 for the data of a small example and the resulting network. The length of the arcs is not an indication of the duration of the tasks, only the connection between two nodes. The objective is to find a path through the network from the starting node to the final node that finishes as early as possible, while all tasks have been completed. This is referred to as the critical path. In our small sample it is a straightforward matter to find that path by considering all possible paths:

Path ABC: $4 + 3 + 3 = 10$
Path DEC: $4 + 1 + 3 = 8$
Path DF: $4 + 3 = 7$

Activity	A	B	C	D	E	F
Duration	4	3	3	4	1	3
Predecessors		A	E, B		D	D

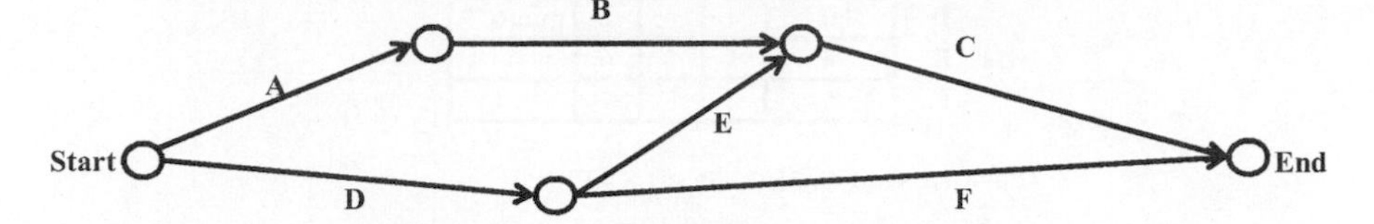

Figure B.1 An Example of Activity on Arc (AOA) Representation of a Project

ABC is called the critical path because if any task along it is delayed, the whole project is also delayed.

Along DEC, either D or E or both may be delayed without affecting the final completion. Note that C cannot start until time 7, when A and B have been completed. D and E, on the other hand, only take 5 time units, and have what is called a slack of 2 time units.

The determination of the critical path in more complex networks is made with an algorithm that makes a forward path through all the tasks from the start to the end and then a backward pass back to the starting node. We need a few more definitions to describe the algorithm:

ES_j—Earliest start of an activity j
EF_j—Earliest finish of an activity j
LS_j—Latest start of an activity j
LF_j—Latest finish of an activity j
d_j—duration of an activity j

Algorithm B.1

1. Calculate the earliest start and finish for each activity. For activities with multiple predecessors, the earliest start is determined by the largest of the earliest finishes of the predecessors.
2. Using the largest EF_j as the LF_j, determine the latest finish and start for each activity. For activities with multiple successors, the latest finish is the minimum of the latest starts of the successors.

The application of the algorithm to our example is shown in Figure B.2. The critical path consists of the activities with zero slack, i.e., ABC = 10.

The alternate representation is called activity on node (AON) and is shown in Figure B.3 for the same example.

Both methods yield the same results. The choice between the two is a matter of taste and prior practice. However, it is usually easier to represent a complex network with the activity on node approach. The representation as an activity on arc requires the labeling of the nodes with an increasing and distinct numerical

	ES	d	EF	LS	LF
A	0	4	4	0	4
B	4	3	7	4	7
C	max(7,5)	3	10	7	10
D	0	4	4	2	min(6,7)
E	4	1	5	6	7
F	5	3	8	7	10

Slack S = LF - EF = LS - ES

$$S_A = 4 - 4 = 0 - 0 = 0$$
$$S_B = 7 - 7 = 4 - 4 = 0$$
$$S_C = 10 - 10 = 7 - 7 = 0$$
$$S_D = 6 - 4 = 2 - 0 = 2$$
$$S_E = 7 - 5 = 6 - 4 = 2$$
$$S_F = 10 - 8 = 7 - 5 = 2$$

Figure B.2 Algorithm B-1 Applied to the Example

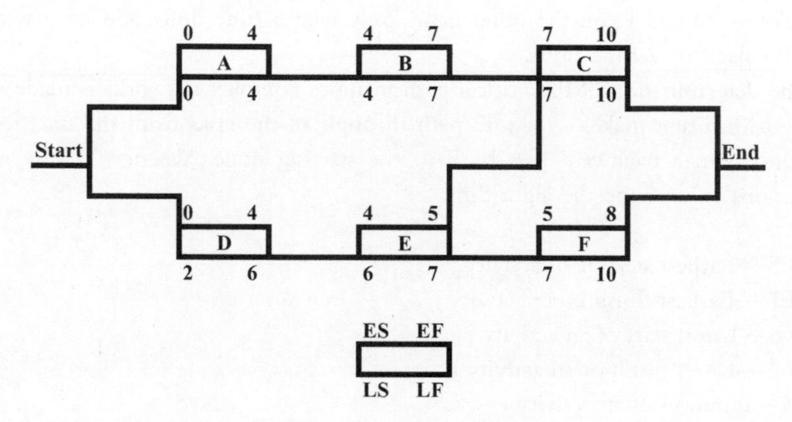

Figure B.3 Activity on Node (AON) Representation of Our Example

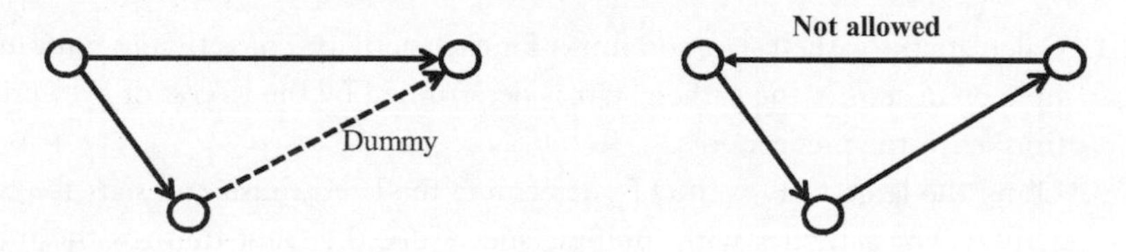

Figure B.4 Examples of a Dummy Activity and Loop Path

integer value (called a topological order) and drawing the network observing the order of the nodes and not entering an activity until all of its predecessors have been entered. A node may not have two arcs entering it. This can be avoided by using dummy activities. Also, we cannot have re-entrant loops in the network (see Figure B.4).

Our objective here is to determine when a project will be completed and assign a probability to that event. In order to be able to do this, we have to estimate how long each task will take. It is natural to ask the opinion of practitioners for these estimates. If the practitioner is relatively certain about the duration of an activity then we do not have to concern ourselves about probabilities. However, it is much more common to get estimates of most optimistic, most pessimistic and most likely durations for a particular activity. This will allow us to calculate three separate estimates for the completion of the project. However, this is not very practical as it is extremely unlikely that all activities will be pessimistic or optimistic, or most likely, for that matter. By assuming triangular distributions for the durations for a second example we arrive at Figure B.5. The average and variance were calculated using the definitions of a triangular distribution:

$$\text{Average} = \frac{(\text{pessimistic} + 4{\star}\text{most likely} + \text{optimistic})}{6} \qquad \text{Equ. B.1}$$

$$\text{Variance} = \left(\frac{\textit{pessimistic} - \textit{optimistic}}{6}\right)^2 \qquad \text{Equ. B.2}$$

Some practitioners will use 4 in the denominator of the variance for greater variability. In the same manner, the average may be calculated by assigning different weights to the components, while ensuring that the denominator is the sum of the coefficients in the numerator.

We can now calculate the length of each path, together with its variance and thus compare probabilities for each path completing at a particular time. We assume that due the central limit theorem the lengths of the paths are normally distributed. Calculating the probabilities with the normal distribution, we can see that despite ACF being the longest average path, because of large variations on path ADEF, the probability of completing at a time is worse with the shorter path. Thus it is not only important to find the longest path, but also important to check the probability of shorter paths.

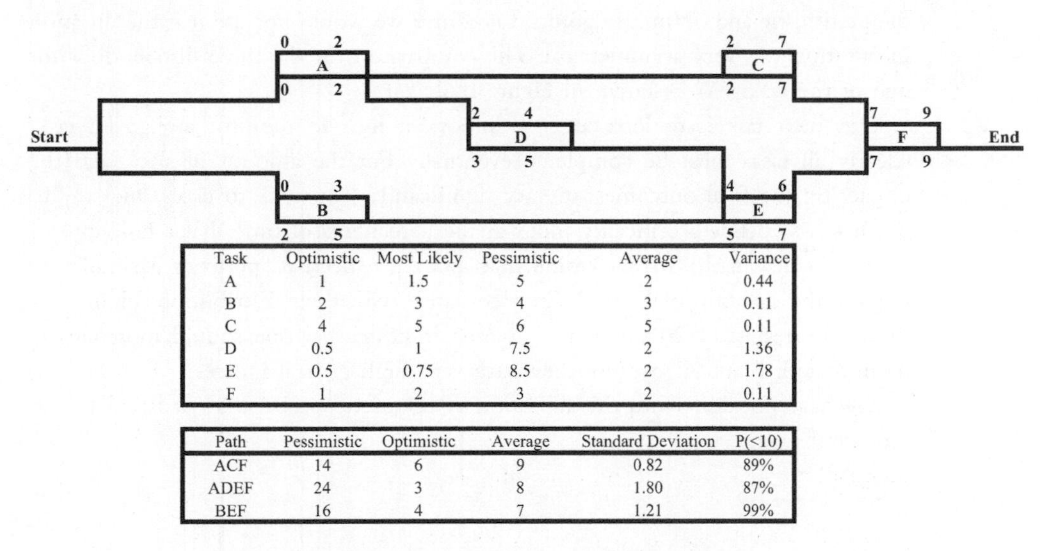

Task	Optimistic	Most Likely	Pessimistic	Average	Variance
A	1	1.5	5	2	0.44
B	2	3	4	3	0.11
C	4	5	6	5	0.11
D	0.5	1	7.5	2	1.36
E	0.5	0.75	8.5	2	1.78
F	1	2	3	2	0.11

Path	Pessimistic	Optimistic	Average	Standard Deviation	P(<10)
ACF	14	6	9	0.82	89%
ADEF	24	3	8	1.80	87%
BEF	16	4	7	1.21	99%

Figure B.5 Probabilistic Values for the Durations of the Second Example

While the triangular distribution is useful for simulating the durations, it has thick tails, accentuating probabilities near the ends of the distribution. The beta distribution is therefore better suited for the simulation. It has fixed ends like the triangular but tends to be shaped like the normal, but does not have to be symmetrical. It has four parameters: minimum, maximum and two shape factors, generally referred to as α and β. The relationship among average and variance and end points a and b are given in the following equations for the average, variance, and mode in terms of the parameters. One can also use Excel's formulas to simulate the beta distribution.

$$\mu = a + (b-a)\left(\frac{\alpha}{\alpha+\beta}\right) \qquad \text{Equ. B.3}$$

$$\sigma^2 = (b-a)^2\left(\frac{\alpha\beta}{(\alpha+\beta)^2(\alpha+\beta+1)}\right) \qquad \text{Equ. B.4}$$

$$m = (b-a)\left(\frac{\alpha-1}{(\alpha+\beta-2)}\right) \qquad \text{Equ. B.5}$$

We can also solve for the parameters:

$$\alpha = \left(\frac{(\beta-2)(m-a)+(b-a)}{b-m}\right) \qquad \text{Equ. B.6}$$

$$\beta = \left(\frac{\alpha(b-m)+2m-(a+b)}{(m-a)}\right) \qquad \text{Equ. B.7}$$

We do have another option and that is to use the normal and truncate the tails at the pessimistic and optimistic points. Of course we would not use it if the duration in question was very asymmetrical. The comparison for the three distributions for one of the tasks (A) is shown in Figure B.6.

Our next task is to look at how important it is to monitor any given task. Clearly all tasks must be completed eventually, but the amount of slack and the impact on the final outcome can vary significantly from task to task. One way to do that is to define a criticality index as the percentage of time that a task appears on the critical path. To determine that index we need to perform a simulation varying the duration of the tasks in accordance with their distribution. Figure B.7 shows the results of 20 simulations, which indicate that one should most closely monitor task F and C because they have very high critical indices.

We have not dealt with the allocation of resources to our task—it is left to our interested readers.

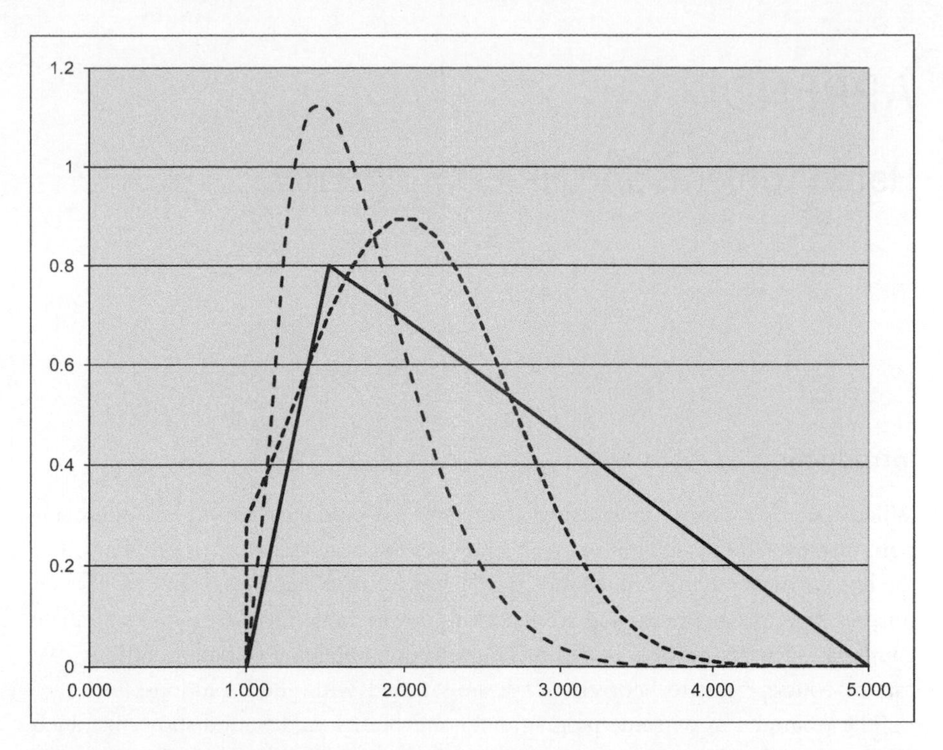

Figure B.6 Comparing Different Distributions to Represent Task Duration. Solid—triangular; heavy dash—normal; dash—beta.

Simulation	Sampled duration of the task							Length of path			
	A	B	C	D	E	F		ACF	ADEF	BEF	Critical Path
1	1.91	3.45	4.87	2.05	3.49	2.31		9.09	9.76	9.25	ADEF
2	1.27	2.67	4.51	1.56	4.35	1.68		7.46	8.86	8.7	ADEF
3	1.8	2.09	5.11	2.33	2.19	2.14		9.05	8.46	6.42	ACF
4	1.24	3.05	5.04	2.30	3.48	1.87		8.15	8.89	8.4	ADEF
5	1.58	2.68	5.22	1.77	0.76	2.60		9.40	6.71	6.04	ACF
6	1.5	2.78	4.68	2.14	2.07	1.46		7.64	7.17	6.31	ACF
7	1.33	2.83	5.61	2.48	5.30	2.48		9.42	11.59	10.61	ADEF
8	1.57	3.3	4.82	3.10	3.30	1.93		8.32	9.90	8.53	ADEF
9	1.5	3.49	4.69	2.05	2.03	1.94		8.13	7.52	7.46	ACF
10	1.87	2.6	5.49	3.27	1.31	2.48		9.84	8.93	6.39	ACF
11	1.75	3.59	4.66	4.69	1.66	1.92		8.33	10.02	7.17	ADEF
12	3.01	2.78	5.06	3.84	2.33	1.65		9.72	10.83	6.76	ADEF
13	1.33	2.36	4.9	2.62	2.37	2.17		8.40	8.49	6.9	ADEF
14	2.14	3.55	5.14	0.55	3.84	2.00		9.28	8.53	9.39	BEF
15	1.25	2.8	5.23	1.92	1.03	1.98		8.46	6.18	5.81	ACF
16	2.17	2.77	5.13	1.97	0.52	1.69		8.99	6.35	4.98	ACF
17	1.41	3.12	5.13	2.55	2.87	1.51		8.05	8.34	7.5	ADEF
18	1.73	2.55	4.92	3.43	3.75	1.45		8.10	10.36	7.75	ADEF
19	1.32	3.46	4.47	1.36	0.82	2.07		7.86	5.57	6.35	ACF
20	1.31	3.51	5.32	6.59	2.20	1.76		8.39	11.86	7.47	ADEF
No. times on Critical Path	19	1	8	11	11	20	Average	8.60	8.72	7.41	
Critical Index	0.95	0.05	0.4	0.55	0.55	1	Std. Dev.	0.69	1.77	1.40	
	Parameters										
Alpha	3	4	4	1	1	4					
Beta	20	4	4	3	4	4					
Min	1	2	4	0.5	0.5	1					
Max	5	4	6	7.5	8.5	3					

Figure B.7 Simulating the Second Example

APPENDIX C

Hard Problems and NP-Completeness

Introduction

When there is a choice, constructive algorithms are preferable to implicit enumeration. However, there are constructive techniques for only the simplest problems. For the rest we must use implicit enumeration unless we find ourselves unable to develop suitable dynamic programming recursions or lower bounds; in that case complete enumeration is unavoidable if we limit ourselves to obtaining optimal solutions. We can, of course, resort to heuristics if we are satisfied with good solutions.

The examples in dynamic programming and branch and bound show that such solutions can involve a lot of computation. We can quote an empirical observation on its average performance. Ignall and Schrage (1965) report that their algorithm for the $n/3/F/C_{max}$ problem requires on average about twice as much time for $(n + 1)$ jobs as it does for n. Thus, if it takes 1 second to solve an n-job problem, it will take 2^r seconds to solve an $(n + r)$ one. For $r = 20$, 2^r seconds = 12 days.

These formidable computational requirements provide the greatest possible encouragement to use constructive methods whenever possible. We should also devote our efforts to finding constructive algorithms for a wider range of problems. Such efforts are likely to be wasted as we will see in the rest of this appendix. Theory predicts that no constructive algorithms will ever be developed for the majority of scheduling problems.

Assumptions for This Appendix

We make no assumptions to limit the range of problems discussed, but we do henceforth assume that all data (i.e., processing times, due dates) are integers. This does not restrict us because no data are ever known to more than a finite number of decimal places. Multiplying all the data in a problem by 10r for some r will result in an essentially equivalent problem, but one in which all data are integers.

What Is a Good Way of Solving a Problem?

Why do we prefer a constructive solution to an enumerative one? It's not enough just to say because it requires less computation. Everyone would prefer to solve a $4/3/F/C_{max}$ problem by branch and bound than a $1000/1//\bar{C}$ problem by the SPT rule. Clearly we should say that we prefer constructive methods because for a given size of problem they need less computation. Our first objective is to make this statement more precise.

Throughout this text we referred to problems in two distinct ways: first, to describe general classes, e.g., the $n/m/G/\overline{T}$ problem; second, to refer to a particular case with particular data, e.g., the problem faced by Albert and friends. Here we only use problem in the sense of a class of problems. Particular cases will be referred to as instances. We are not concerned with the computation necessary to solve a given size of problem but rather a given size of instance.

So what do we mean by the size of an instance? Computer scientists use the following definition. Consider the description of $6/1//\overline{F}$ instance with processing times

$$p_1 = 10, p_2 = 6, p_3 = 8, p_4 = 16, p_5 = 14, P_6 = 7$$

Given that we know the problem type is $n/1//\overline{F}$, we need only specify the numbers 10, 6, 8, 16, 14, 7 to describe the problem precisely. We can find that $n = 6$ by counting the number in the list and we can find that, say, $p_3 = 8$ by using the natural convention that the ith number is p_i. The commas are essential. The string of digits 106816147, which results when commas are omitted, is meaningless. Of course, we could replace the commas with colons or vertical strokes; but some separator symbol is necessary. In the above string there are 14 digits and separators. Hence we say the instance has size 14. In order to determine the size of an instance we need to know:

1. The type of problem that it is, that is its n/m/A/B classification;
2. The encoding convention used to list the data of the instance.

The definition of the size of an instance depends upon the encoding convention adopted. This dependence on the convention used is irrelevant to our needs, provided that we always adopt a reasonable encoding convention. By reasonable we mean that there are no unnecessary symbols or data and that all the numbers are to the same base. The size of instances is effectively independent of the encoding convention that we use.

The size of an instance is determined both by the number of data needed to define the instance and also by their magnitude. There is a dependence on their number because each quantity requires at least one digit to represent it. Hence the instance size must be greater than the number of data defining it. There is a dependence on the magnitudes of the data because the larger any one of them is the more digits will be required to represent it. But there is a difference between these dependencies. The first is linear; the second is logarithmic. Suppose that we have 10 numbers all of roughly the same magnitude. Then to represent all of them we need about 10 times as many digits as for any one. If we were to consider a hundred such numbers, we would need a hundred times as many digits. However, suppose that we consider two numbers, the first 10 times as large as the second. Then we need only one more digit to represent the first than we do for the second. (This is so if we represent numbers to the base 10; a similar observation can be made for other bases.) If the first number is a hundred times as big, we need only two more digits. Thus, while the size of an instance does depend both on the number of data and on their magnitude, the dependencies are not equal.

We are now in a position to discuss the relative computational merits of different algorithms. We shall do this by means of time complexity functions. The time complexity function f(v) of an algorithm gives the maximum number of operations that would be required to solve an instance of size v.

In practice, it is unnecessary for us to determine time complexity functions completely; all we need is some indication of their behavior as the problem size increases. Because of this, we introduce a very useful notation. We say that f(v) is O(g(v)); read f is of the same order as g whenever

$$f(v)/g(v) \rightarrow c, \text{ some constant, as } v \rightarrow \infty \qquad \text{Equ. C.1}$$

Thus f(v) is O(g(v)) if their ratio tends to become constant as v increases. In other words, as v increases the behavior of f(v) and g(v) become more and more similar until they are essentially the same. The following properties hold (n and m are integers): If f(v) is $O(v^n)$ and g(v) is $O(v^m)$, then:

$f(v)g(v)$ is $O(v^{n+m})$
$f(v)/g(v)$ is $O(v^{n-m})$
$f(g(v))$ is $O(v^{mn})$
The polynomial $(a_n v^n + a_{n-1} v^{n-1} + \ldots + a_0)$ is $O(v^n)$

We shall say that an algorithm has polynomial time complexity if its time complexity function f(v) is O(p(v)) for some polynomial p(v). Otherwise an algorithm has exponential time complexity. Note that we include in our definition of exponential time complexity behavior that is not normally regarded as exponential. For instance, if f(v) is O(v!) we say that it exhibits exponential behavior.

The reason for making these definitions becomes apparent when we examine Figures C.1 and C.2. While computers have increased in speed many times over

Time Complexity function	v					
	10	20	30	40	50	60
v	0.00001 sec	0.00002 sec	0.00003 sec	0.00004 sec	0.00005 sec	0.00006 sec
v^2	0.0001 sec	0.01104 sec	0.0009 sec	0.0016 sec	0.0025 sec	0.0036 sec
v^5	0.1 sec	3.2 sec	24.3 sec	1.7 min	5.2 min	13 min
v^{10}	2.7 hrs	118.5 days	18.7 yrs.	3.3 centuries	30.9 centuries	192 centuries
2^v	0.001 sec	1.0 sec	17.9 min	12.7 days	35.7 yrs	366 centuries
3^v	0.59 sec	58 min	6.5 yrs	3855 centuries	$2\text{x}10^8$ centuries	$1.3\text{x}10^{13}$ centuries
$v!$	3.6 sec	770 centuries	$8.4\text{x}10^{16}$ centuries	$2.5\text{x}10^{32}$ centuries	$9.6\text{x}10^{48}$ centuries	$2.6\text{x}19^{66}$ centuries

Figure C.1 The Time Requirements of Algorithms with Certain Time Complexity Functions under the Assumption That One Mathematical Operation Takes One Microsecond
Source: French, 1982.

Time Complexity Function	Size of instance solved in a given time on slow computer	Size of instance solved in the same time on a computer 1000 times faster
v	v_1	$1000v_1$
v^2	v_2	$31.62v_2$
v^5	v_3	$3.98v_3$
v	v_4	$1.99v_4$
2^v	v_5	$v_5 + 10$
3^v	v_6	$v_6 + 6$
$v!$	v_7	$v_7 + 3 \quad v_7 \leq 10$ $v_7 + 2 \quad 10 < v_7 \leq 30$ $v_7 + 1 \quad 30 < v_7 \leq 1000$

Figure C.2 Increase in Instance Size Solvable in a Given Time for a Thousand-Fold Increase in Computing Speed

Source: French, 1982.

since these were published, the ratios remain the same. Figure C.1 compares the actual time requirements of several polynomial and exponential time complexity functions.

The first four rows of this figure illustrate the time requirements for algorithms exhibiting polynomial behavior; the last three illustrate exponential behavior. It is apparent that the increase in time requirements with instance size is far less dramatic for the polynomially bounded algorithms than for the others. Empirically the distinction between polynomially and exponentially bounded algorithms is very marked.

It is also informative to consider how much larger an instance can be solved in a given time for a given increase in computing speed. To make this question exact, we compare the performance of an algorithm on two computers. The slower one takes a microsecond to perform a mathematical operation; the faster one takes a nanosecond. If in a given time we can solve an instance of size v on the first computer, how much larger an instance can we solve in the same time on the second? The answer is given in Figure C.2.

It can be seen that algorithms with polynomial time complexity allow a multiplicative increase in instance size for a given gain in computer power; whereas those with exponential time complexity only allow an additive increase in instance size. If we have algorithms with polynomial time complexity, we know that advances in computer technology will enable us to solve much larger instances. However, if we only have solution techniques that are of exponential time complexity, we know that the gains from increased computing power will be slight.

For these reasons it has become conventional to talk of problems as being well solved when we have developed an algorithm with polynomial time complexity. We are further encouraged in this definition because the existence of a polynomial time algorithm usually means that we have solved the problem through some insight, some understanding of the structure of solution. Exponential time algorithms are mainly types of exhaustive search procedures. The logic of the search may be very subtle and clever, but it seldom shows that we really understand the problem that we are trying to solve. You may appreciate this point from consideration of our earlier work. The constructive algorithms of Chapters 8 and 9 are of polynomial time complexity, whereas dynamic programming and branch and bound algorithms have exponential time complexity. Perhaps we had better examine these assertions more closely.

As an example, let's confirm that there is a polynomial time algorithm for constructing an SPT sequence in an $n/1//\bar{F}$ problem. Our aim is simply to sort the jobs into order of non-decreasing processing times. A straightforward way of doing this is as follows. We begin by listing the jobs in any order; for convenience we take this to be $(J_1, J_2, \ldots, J_n)$, We pass through the sequence comparing adjacent jobs, beginning with the jobs in the first and second positions. If their processing times are in decreasing order, i.e., if the former has the longer processing time, we interchange them. Otherwise, we leave them in their current positions. Next we compare the jobs in the second and third positions. (The job in the second position may have been placed there as a result of the first comparison and interchange.) If the processing times are in decreasing order, we interchange the second and third jobs. Otherwise we leave them. We continue through the sequence, comparing adjacent jobs. Thus we shall make (n – 1) comparisons and at most (n – 1) interchanges with the result that the longest job is moved down the sequence to the last position. For example, consider a $5/1//\bar{F}$ instance with data: $p_1 = 6$, $p_2 = 3$, $p_3 = 7$, $p_4 = 1$, and $p_5 = 3$.

Applying this method we would make the interchanges shown in the first four rows of Figure C.3. Note that the longest job, J_3, is in the last position of the sequences after the four comparisons and appropriate interchanges.

Next we pass through the current sequence again, interchanging jobs in the same fashion. Doing so guarantees that we will have moved the job with the second longest processing time into the (n – l)st position of the sequence. Because we have already ensured that the last position is occupied by a job with the longest processing time, we need only compare and, perhaps, interchange the (n – 2) pairs fanned from the first (n – 1) jobs in the sequence. These comparisons correspond with

Current Sequence	Compare positions	Results
(1,2,3,4,5)	1st and 2nd	$p_1 = 6 > 3 = p_2$ so exchange
(2,1,3,4,5)	2nd and 3rd	$p_1 = 6 < 7 = p_3$ so leave
(2,1,3,4,5)	3rd and 4th	$p_3 = 7 > 1 = p_4$ so exchange
(2,1,4,3,5)	4th and 5th	$p_3 = 7 > 3 = p_5$ so exchange
(2,1,4,5,3)	1st and 2nd	$p_2 = 3 < 6 = p_1$ so leave
(2,1,4,5,3)	2nd and 3rd	$p_1 = 6 > 1 = p_4$ so exchange
(2,4,1,5,3)	3rd and 4th	$p_1 = 6 > 3 = p_5$ so exchange
(2,4,5,1,3)	1st and 2nd	$p_2 = 3 > 6 = p_4$ so exchange
(4,2,5,1,3)	2nd and 3rd	$P_3 = 3 = 3 = p_5$ so leave
(4,2,5,1,3)	1st and 2nd	$P_4 = 1 < 3 = p_2$ so leave
(4,2,5,1,3)		

Figure C.3 Sorting the Jobs into SPT Order for Our Particular Instance

those in lines 5, 6, and 7 of Figure C-3. We continue this procedure, passing through the current sequence again and again, interchanging jobs so that during the rth pass the rth longest job is put into the (n − r + l)st position. Ultimately we find an SPT sequence. Obviously, whenever two jobs have the same processing time their order is immaterial and we do not interchange such jobs.

The pass that locates the rth longest job in the (n − r + 1)st position requires (n − r) comparisons and at most the same number of interchanges. Thus there will be

$$\sum_{r=1}^{n}(n-r)=\frac{n}{2}(n-1)$$

comparisons and perhaps the same number of interchanges. Moreover, roughly the same amount of work will be required to increment and check the loop controls. Thus this algorithm will require $O(n(n-1)/2) = O(n^2)$ operations to construct the SPT sequence. Any encoding convention will require at least one digit for each p_i. Hence n cannot exceed v, the instance size. It follows that the time complexity of the algorithm is almost $O(v^2)$ and therefore, it is a polynomial time algorithm. (The sort procedure that we have used is by no means the most efficient available). However, our simple algorithm is all we need to show that a polynomial time algorithm exists.)

By similar arguments it may also be shown that all the constructive algorithms discussed earlier have polynomial time complexity.

French (1982) determined that a dynamic programming solution to the $n/1//\overline{T}$ problem required $O(n2^{n-1})$ operations and that complete enumeration required $6n(n!) + 3(n! - 1)$. This implies exponential time behavior as you can see in Figure C.4. We can confirm this behavior theoretically. To specify an instance we need 2n numbers, namely $p_1, p_2, \ldots, p_n$ and $d_1, d_2, \ldots, d_n$. We shall also need $2n - 1$ separators. Thus, if we construct an example in which each number may be represented by exactly k digits, we obtain an instance of size: $v = 2kn + 2n - 1$ and solving for n:

$$n=\frac{v+1}{2(k+1)}$$

Thus this instance will take $O\left(\frac{v+1}{2(k+1)}2^{\left(\frac{v+1}{2(k+1)}-1\right)}\right)$ operations to solve, which implies exponential time complexity.

In conclusion we say that a scheduling problem has been well solved if we have found an algorithm with polynomial time complexity. Such algorithms are usually

n	Number of operations required by Complete enumeration	Dynamic programming
4	647	237
10	2.29×10^8	33789
20	2.99×10^{20}	6.40×10^7
40	1.98×10^{50}	1.35×10^{14}

Figure C.4 Comparison of the Computation Required to Solve an $n/1//\overline{T}$ Problem

constructive in nature and imply an understanding of the structure of the solution. On the other hand, dynamic programming and branch and bound solutions exhibit exponential time behavior.

The Classes P and NP

The class P consists of all problems for which algorithms with polynomial time behavior have been found. The class NP is essentially the set of problems for which algorithms with exponential behavior have been found. P is contained in NP: if we have a polynomial time algorithm for a problem, we can always inflate it inefficiently so that it takes exponential time. Occasionally a problem originally in NP but not in P is moved into P, as someone discovers a polynomial time algorithm. Unfortunately, it is one of the most widely held conjectures of modern mathematics that there are problems in NP that may never be moved into P. In other words, many mathematicians believe that algorithms with polynomial time complexity will never be found for certain problems, essentially because they are just too hard. Most scheduling problems fall into this category.

We note that the classes P and NP are not confined to scheduling, and the concepts apply to problems arising throughout combinatorial theory.

We now turn to the discussion of the conjecture that $P \neq NP$ and its implications for scheduling problems. A central concept in the argument is the idea of reducing one problem to another. To introduce this through an example, consider the discussion in Chapter 11 on integer programming. There it was shown that a certain scheduling problem could be translated into an equivalent integer programming problem, i.e., we reduced our scheduling problem to an integer program.

Suppose that we can reduce one problem Π_1 to another Π_2 in polynomial time; then we say Π_1 is polynomially reducible to problem Π_2 and we write $\Pi_1 \rightarrow \Pi_2$. We say that a problem Π lying in NP is NP-complete if every other problem in NP is polynomially reducible to Π.

Thus the NP-complete problems form a subclass of NP. Furthermore, this subclass is formed of the hardest problem in NP. For if we find a polynomial time algorithm for any NP-complete problem, then we can answer all problems in NP in polynomial time. Thus if one such algorithm is found, $P = NP$. The conjecture that $P \neq NP$ is, therefore, equivalent to the conjecture that of NP-complete problem can be answered in polynomial time.

In fact, most people who have worked in the area have become convinced that NP-complete problems cannot be solved in polynomial time. This conviction arises not just because it is hard to admit failure, but familiarity with NP-complete problems brings a deep appreciation of their very great difficulty—one does not know how hard a problem is until one has tried it. Despite all this empirical evidence, it cannot be repeated too often that $P \neq NP$ remains a conjecture.

And now on the point of this appendix. It has been shown that very many scheduling problems are NP-complete. For instance, both the $n/1//\sum\gamma_i$ and the general $n/3/F/C_{max}$ problems are NP-complete. Hence; for the present there is no alternative to trying to solve these by implicit enumeration, as we did with dynamic programming and branch and bound. The introduction of non-zero ready times usually renders a problem NP-complete; thus the $n/1//\overline{C}$ and $n/1//L_{max}$ problems all became NP-complete when non-zero ready times are allowed. The $n/m/F/C_{max}$ problem is NP-complete for all $m \geq 3$ and so naturally is the $n/m/G/C_{max}$ problem. Thus it is not at all surprising that constructive algorithms have not been found for these problems.

APPENDIX D

Problems

Problems are arranged by chapter but not necessarily by degree of difficulty. All problems have been used in previous examinations and assignments

D1.1 Our newspaper reading friends (from Chapter 1) have decided that they want to be sure about when they leave for the walk in the park and it should be as early as possible. One of them, Daniel, has moved away, so there are only three of them. They have, however, decided to expand their horizons, and because they can all read French, they have added *Le Monde* to their list of papers. And wanting some opinions different from the *Financial Times*, also added the *Wall Street Journal*. In the interest of simplicity, they will forgo their favorite reading order and now are willing to each read the papers in the same order (F,L,E,N,LM,WSJ). Figure D1.1 gives the amount of time each paper takes to be read by each person in minutes. They all start reading at 8:00 A.M. When is the earliest that they can leave for the park? Second, is it likely that there is more than one schedule that can accomplish this time? About how many do you think and why?

D3.1 Assume that a big box store buys a particular television set for $200. The cost of placing an order for a shipment of TVs is $35. The cost of keeping inventory is 20% per year. They expect to sell an average of 900 TV sets in a year (standard deviation 40), at a fairly steady rate. The lead time is 2 weeks. How much safety stock should they keep in order to stock out less than 5% of order cycles if they used an (s,S) system? An (R,S) system? Also specify S and R. Redo for a service level of 95%.

D3.2 Assume that Best Buy buys a particular digital camera for $100. The cost of placing an order (including shipping) of cameras is $70. The cost of keeping inventory for Best Buy is 24% per year. They expect to sell an average of 500 cameras of this type in a year (standard deviation 40), at a fairly steady rate. The lead time is 2 weeks.

Calculate Q, s, and R without safety stock. How much safety stock should they keep in order to stock out less than 5% of order cycles if they used an (s,Q) system? How much safety stock should they keep in order to maintain a 99% service level if they used an (R,S) system?

D3.3 Determine the order point, quantity, and safety stock (90% of ordering cycles without a shortage) for the following data. Ordering cost is $25. Annual demand is 8,000. Annual inventory carrying rate is 24%. Value of the product is $100. Variance of the annual demand is 10,000. Lead time is 2 months.

Job	F	L	E	N	LM	WJS
A	10	12	8	45	10	5
B	5	3	4	5	5	5
C	20	4	4	2	30	10

Figure D1.1 Reading Order and Duration

D3.4 The lead time for an item is 2 months. The average demand is 36 per year, standard deviation = 2. Carrying cost is 20%, the value of the item is $95,000, and the ordering cost is $320. What is the order quantity? How often should you order? How much safety stock should be kept if the goal is to stock out no more than in one order cycle per year?

D3.5 A retailer wishes to use an (R,s,S) system with a cost of a stockout of $50. The other relevant information is: annual demand 1,700 with a standard deviation of 50, ordering cost of $55, a lead time of 1 month, inventory carrying cost of 15%, and value of $100. Determine a reasonable set of values for (R,s,S) and compute the cost for the first four cycles, starting with maximum inventory.

D4.1 Define the variables and relevant costs (symbols only—not amounts) and write the equations for the following problem to minimize total costs:

> A company has 50 products and plans monthly for a year. The demand is known for each product (D_{it}) in each month t. You can hire both permanent (HP) and temporary employees (HT) at a different cost, although their regular pay is the same. However, temporary workers may not work overtime and must be let go after 3 months. Permanent employees may be terminated only after being with the company for 6 months. Other ways of increasing capacity are the use inventory, but not shortages. Overtime is subject to a maximum. We know the hours K_i required to produce each product and the number of hours per employee available each month.

D4.2 A family of products at a specialty table manufacturing company has the following projected monthly demand.

Period	1	2	3	4	5	6
Demand	1000	850	600	500	650	700

Figure D4.2 Monthly Demand

There are currently 30 workers who can each produce 50 tables per month. Inventory and subcontracting are not feasible. Workers may be laid off at $5,000 each. Hiring a worker costs $4,000. Workers are paid whether they are making tables or not. The shortage cost is estimated at $300. The company faces a decision whether to conduct a promotional campaign that would cost $35,000 and increase the demand in each of the last 3 months to 750 units. Material and overhead cost per table is $600. Workers get paid $4,500 per month. The sales price before the promotion is $1,000 per table.

What is the total profit when you optimize hiring and firing and production before the promotion? After the promotion? So would you do it? The price is the same for all periods.

What price after the promotion would make the two scenarios equal, assuming that the new price is applicable for all periods?

D4.3 There are 12 periods, for which the demand is given. Two hours of labor for each unit of product. You start with 30 people whom you can hire or fire at $5,000 per person. Regular pay is $19 (you only pay labor for units you produce). Overtime is $25/hour. Shortage cost is $220 per unit; inventory is $20 per unit/month. All demand must be satisfied by the end of the 12 periods. Overtime is limited to 600 hours per month. There are 160 hours in a month. Find the minimum cost.

Month	1	2	3	4	5	6	7	8	9	10	11	12
Demand	2,400	3,500	3,000	3,900	2,100	2,400	3,700	4,000	3,500	3,000	3,300	2,800

Figure D4.3 Demand Data

D4.4 A factory has forecast demand for each of their two products for the next 12 months that they may meet but cannot exceed. They currently have 200 employees who are on salary ($3,800/month), that is they have to be paid whether they are producing product or not. There are 160 productive hours per worker in a month. Overtime is on an hourly basis and costs $36/hour. Maximum overtime per employee per month is 40 hours. Employees may be hired ($5,000 each), but not laid off during this year. The selling price of the two products is $350 and $45 respectively. The amount of time required to produce each product is 5 hours and 1 hour respectively. Ignore material and overhead costs. Inventory carrying costs are $7 and $0.90 per month respectively. Shortages are not allowed. Initial inventories are zero. Maximize the profit.

Period	Demand 1	Demand 2	Period	Demand 1	Demand 2
1	2,533	17,869	7	2,528	22,609
2	2,460	19,175	8	2,455	21,414
3	2,072	20,560	9	2,696	24,246
4	2,289	20,872	10	2,915	23,610
5	2,732	22,054	11	2,486	21,443
6	2,760	21,964	12	2,675	20,315

Figure D4.4 Demand Data for Two Products

D5.1 The orders and forecasts for a product are given in Figure D5.1 for 6 months. The usual order quantity is 200 units. The lead time and the demand time fence are both 2 months. The ordering cost is $50 and the inventory carrying cost per period is $1.50. (a) Complete the MRP record, including the available to promise. (b) How much money would you save if instead ordering the 200 units when you needed to order, you ordered exactly what you needed?

	Initial Inventory	1	2	3	4	5	6
Forecast		100	100	120	150	140	120
Orders		90	110	125	95	85	75
POH	350						

Figure D5.1 MRP Record

(c) After completing the MRP record in (a), if a customer asked for an additional 20 units in period 2, what would you tell the customer and what action, if any, would you take?

D5.2 We have two products with two level bills of materials and routings through two work centers. Given that information as well as the demand and processing time, determine the load in each work center in all necessary periods (8-hour, 5-day weeks). Do not perform work earlier than necessary.

Product A consists of 1 item C and 2 items D
Product B consists of 1 Item D
Routing A and B—Workcenter X
Routing C and D—Workcenter Y
Demand in period 3–15 A's and 20 B's (due at the end of the period)
Demand in period 4–20 A's and 25 B's (due at the end of the period)

Time per unit

A in X: 1/2 hour
B in X: 1 hour
C in Y: 1 hour
D in Y: 1/2 hour

D5.3 Create a bill of materials for two products to which you have access or know about and that share some components. Create demand files from a number of customers for these two products. List work centers, routings, processing times and set up times, orders, and due dates. Use Access or Excel or other database or spreadsheet software.

D5.4 Create the MRP records for components B, C, D, and E to answer the four questions. Item A is the finished product, made from one B and two Cs. Item B is made from three Ds and one E. Item C is made from one D and two Es. The following are the MPS quantities for item A (MPS quantities, not customer demand).

Period	1	2	3	4	5	6	7	8	9	10	11	12
MPS A			20	20		50		60	30		40	
Item	B	C	D	E								
Lot Size	Lot for lot	150	250	400								
Lead Time	1	2	1	2								
On hand	10	50	10	320								
Scheduled Receipts	None	None	250 wk1	None								
Safety stock	None	None	20	None								

Figure D5.4 Dependent MRP Record

1. Is there anything of special note that is a concern? What and why?
2. The R&D department wants to take the 10 units of D that are on hand immediately and also wants 20 more in week 1. What would you tell them and why?
3. The master scheduler asks you if they can move the MPS of 20 in week 3 to week 2. From the data on the MRP records, what would you tell them and why?
4. R&D has a new design they want to implement for part C, and they are asking you when they should plan to do that. What do you tell them and why?

D6.1 Define mean utilization of the machines as the average proportion of the makespan (C_{max}) for which the machines are actually processing. Show that maximizing mean utilization is equivalent to minimizing the makespan.

D6.2 Create a feasible schedule for the following five jobs and calculate the measures C, T, and L for each job and the associated averages. All jobs are ready at time 0.

Job	1	2	3	4	5
p_{i1}	1	2	3	2	3
p_{i2}	2	3	1	3	1
p_{i3}	3	1	2	2	1
TC	1,2,3	1,3,2	3,2,1	2,1,3	3,2,1
Time due	10	10	6	12	8

Figure D6.2 Processing Times and Technological Constraints

D6.3 Show that $\bar{C} / C_{max}$ is not a regular measure of performance by considering the $2/2/G/\bar{C} / C_{max}$ example with data:

	Technological Constraints and	
Job	1st Machine	2nd Machine
J_1	M_2: $p_{12} = 1$	M_1: $p_{11} = 1$
J_2	M_1: $P_{21} = 1$	M_2: $p_{22} = 1$

Figure D6.3 Data for an Irregular Measure

D6.4 Our newspaper reading friends (from Chapter 1) have decided that they want to be sure about when they leave for the walk in the park and it should be as early as possible. One of them, Daniel, has moved away, so there are only three of them. They have, however, decided to expand their horizons, and because they can all read French, they have added *Le Monde* to their list of papers. And wanting some opinions different from the *Financial Times*, also added the *Wall Street Journal.* In the interest of simplicity, they will forgo their favorite reading order and now are willing to each read the papers in the same order (F,L,E,N,LM,WSJ). Figure D6.4 gives the amount of time each paper takes to be read by each person in minutes. They all start reading at 8:00 A.M. When is the earliest that they can leave for the park? Second, is it likely that there is more than one schedule that can accomplish this time? About how many do you think and why?

Job	F	L	E	N	LM	WJS
A	10	12	8	45	10	5
B	5	3	4	5	5	5
C	20	4	4	2	30	10

Figure D6.4 Reading Order and Times

D6.5 Define mean utilization of the machines as the average proportion of the makespan (C_{max}) for which the machines are actually processing. Show that maximizing mean utilization is equivalent to minimizing the makespan.

D7.1 Apply both the active and non-delay methods to the problem below, starting from time 0. Note that some of these jobs have ready times, so they cannot start at 0. Choose your own objective function and method of breaking ties and explain why you chose them. Calculate C_{max}, T_{max}, $\overline{T}$, $\overline{E}$, and $\overline{F}$ for both schedules and discuss the differences you got from the active and non-delay schedules. What is your overall assessment of this situation?

	J1	J2	J3	J4	J5
r_i	0	2	3	2	15
d_i	16	20	18	25	24
p_{i1}	5	6	4	6	3
p_{i2}	3	3	2	2	2
p_{i3}	2	3		4	
p_{i4}	2	5	2	6	1
TC	1234	1234	124	1234	142

Figure D7.1 Processing Times and Technological Constraints

D7.2 If you were preparing a traditional MRP plan for problem D7.1 with the given due dates, processing times and TC, how would that change the problem? What would you recommend for future orders? (Note that ready times would not be given and would have to be obtained from the due dates by subtracting a multiple of the total processing time). What multiple would you recommend for this problem? Simulate the process backwards from the due dates, using inverse dispatching rules of your choice. What conclusions can you draw from this?

D7.3 Choose a performance measure. Using the data in Figure D7.3, generate an active schedule, draw the Gantt chart, and evaluate your performance measure.

Job	p_{i1}	p_{i2}	p_{i3}	p_{i4}	d	TC
1	2	3	4	2	11	1234
2	3	2	4	1	13	1234
3	4	-	3	2	17	134
4	3	-	1	1	9	134
5	5	3	3	2	18	1243
6	1	1	2	3	9	1243
7	2	-	2	-	12	31
8	-	3	3	6	15	234
9	-	4	4	3	19	234
10	3	2	1	3	12	2134

Figure D7.3 Processing Times and Technological Constraints

Make whatever selections you want when the choice is arbitrary. Discuss your results. Repeat for a non-delay schedule.

D8.1 Solve the $8/1//L_{max}$ with precedence constraints 3 before 6 before 1; 4 before 5.

Job	J1	J2	J3	J4	J5	J6	J7	J8
Processing Time	8	4	7	8	4	8	8	3
Due date	21	6	21	6	13	21	15	10

Figure D8.1 Processing Times and Due Dates

D8.2 Determine the smallest maximum penalty and the associated sequence for the parameters given in Figure D8.2.

Job	Processing time	Due date	Penalty formula
1	6	8	Tardiness
2	5	7	1/2(Tardiness + Flow time)
3	3	13	Flow time

Figure D8.2 Different Penalties

D8.3 You are the manager of a machine shop, and need to schedule a number of jobs on your largest Numerical Control (NC) machine. You decide to restrict the order of your choices somewhat to take advantage of subsequent processing that some jobs have to undergo. There are four jobs. Job 3 should be done before job 1. The processing times, due dates and the cost of being late (in terms of time tardy and/or flow time) for each job are given below. Find the sequence that will minimize the maximum cost. What is the total cost of your schedule?

Job	1	2	3	4
p	3	3	4	2
d	8	7	7	10
Cost	2T	F	3T	F/2

Figure D8.3 Different Penalties, Four Jobs

D8.4 Use Smith's algorithm to solve the $7/1//\overline{F}$ subject to $T_{max} \leq 3$ problem with the data in Figure D8.4.

D8.5 Generate all schedules efficient with respect to T_{max} and $\overline{F}$ for the seven-job problem in D8–4. Find the minimum total cost schedule for $c(T_{max}, \overline{F}\}= 3Tmax + 5\overline{F}$.

Job	J1	J2	J3	J4	J5	J6	J7
Processing Time	6	2	4	9	3	1	8
Due Date	33	13	6	22	31	38	14

Figure D8.4 Processing Times and Due Dates, Seven Jobs

D8.6 Solve the $9/1//n_T$ problem and decide what to do with the tardy jobs and explain your choice.

Job	1	2	3	4	5	6	7	8	9
Due Date	19	16	25	3	9	8	29	23	2
Processing Time	5	3	1	2	4	4	2	7	1

Figure D8.6 Processing Times and Due Dates, Nine Jobs

D8.7 You are asked to solve the problem $n/1//\sum_i^n \alpha_i F_i = 1$, where $\sum_{i=1}^n \alpha_i = 1$ and $\alpha_i > 0$ for i = 1, 2, . . ., n, i.e., you have to minimize a weighted mean of the flow times. Show that an optimal schedule sequences the jobs such $\frac{p_{i1}}{\alpha_{i(1)}} \leq \frac{p_{i2}}{\alpha_{i(2)}} \leq \frac{p_{i3}}{\alpha_{i(3)}} \cdots \frac{p_{i1n}}{\alpha_{i(n)}}$. What would the optimal sequence be if some of the $\alpha_i = 0$?

D9.1 Find a good schedule and the corresponding C_{max} for this $7/3/P/F_{max}$ problem. The technological constraints are M1, M2, M3. How good do you think this schedule is? Why? Why may it not be optimal?

Job	1	2	3	4	5	6	7
p_1	6	5	8	6	2	4	2
P_2	3	2	3	1	2	4	2
P_3	3	5	2	3	2	2	2

Figure D9.1 Three Machines, Seven Jobs, Processing Times

D9.2 You have a $6/3/P/F_{max}$ problem with zero ready times. You are to find a solution within 6% of the optimum.

Job	a_i	b_i	c_i
1	7	9	5
2	11	8	3
3	6	7	11
4	4	8	6
5	7	5	9
6	11	8	8

Figure D9.2 Data for a Suboptimal Solution

D10.1 Given the $5/1/P/\bar{T}$ problem, where job 3 must occur before job 2, determine an optimum sequence and the associated $\bar{T}$.

J	1	2	3	4	5
p	3	2	4	8	7
d	9	11	15	27	8

Figure D10.1 Data for Precedences

D10.2 Derive the bounds equations for an $n/5/P/C_{max}$ to be used for the branch and bound method. Then apply it to the initial frontier of the problem given in Figure D10.2.

Job	a_i	b_i	c_i	d_i	e_i
1	6	6	5	5	8
2	2	8	1	10	1
3	1	10	1	5	3
4	2	2	5	10	9

Figure D10.2 Data for the Five Machine Bounds Problem

D10.3 Determine the minimum average tardiness and the corresponding sequence for the given four-job problem. Job 4 must occur after Job 2 in the sequence. Give your reasons for your process.

Job	1	2	3	4
Processing time	3	8	6	7
Due date	10	11	12	17

Figure D10.3 Processing Times and Due Dates

D10.4 Obtain an optimum solution to the $5/1/P/\sum T$ problem in Figure D10.4.

Job		1	2	3	4	5
p		3	8	4	5	6
d		10	7	8	9	17

Figure D10.4 Data for the Tsum Problem

D10.5 This is from an $n/m/P/C_{max}$ branch and bound procedure. The current seed is 100. The frontier search at this node has yielded the following lower bounds:

123XXXXX	124XXXXX	125XXXXX	126XXXXX	127XXXXX	128XXXXX
102	101	98	97	100	107

How many potential schedules are eliminated at this node?

D10.6 Solve the $6/1//\bar{T}$ problem with precedence constraints that J1 must be processed before J2 and J3 and that J4 must be processed immediately before J5, which in turn must be processed immediately before J6. The processing times and due dates of the jobs are in Figure D10.6.

J_i	J_1	J_2	J_3	J_4	J_5	J_6
p_i	6	4	8	2	10	3
d_i	9	12	14	8	20	19

Figure D10.6 Six Jobs Tbar Problem

D11.1 Reformulate the $3/3/P/C_{max}$ problem in Figure D11.1 as an integer program and solve using Excel's Solver.

Processing Times			
Job	M1	M2	M3
1	6	8	3
2	9	5	2
3	4	8	17

Figure D11.1 3/3/P/Cmax Data

D11.2 Show that the $n/m/P/\overline{F}$ problem may be reformulated as an integer program by defining constraints as in Chapter 11 and with the objective function. $\sum_{K=1}^{n}\left(\sum_{J=1}^{m-1} W_{jk} + \sum_{k=1}^{K}\sum_{i=1}^{n} X_{ik} p_{il}\right)$.

D11.3 Show that minimizing the maximum completion in an $n/m/P/C_{max}$ problem is equivalent to minimizing the idle time on the last machine.

D12.1 Perform three steps of simulated annealing with k/K = 0.4 for the 6/1// $\sum T$ problem shown below. Explain your choice of the seed schedule. (Minimum population size is 4.)

J	1	2	3	4	5	6
p	3	4	9	6	5	7
d	14	32	25	11	16	9

Figure D12.1 Data for Simulated Annealing

D12.2 This is a three-job, three-machine general job shop problem with the technological constraints and processing times given in the two tables and an objective to minimize C_{max}. Use the shifting bottleneck heuristic and longest lead time to find a good schedule and C_{max}.

	Processing times			Technological Constraints		
Jobs	**1**	**2**	**3**	**J1**	**J2**	**J3**
p1	**5**	**4**	**1**	**1**	**1**	**2**
p2	**4**	**2**	**4**	**2**	**2**	**3**
p3	**2**	**4**	**3**	**3**	**3**	**1**

Figure D12.2 Processing Times and Technological Constraints

D12.3 Use four or more steps to begin obtaining a good solution for the $8/1//\overline{T}$ problem shown in Figure D12.3. What was your seed and why did you choose it?

J	1	2	3	4	5	6	7	8
p	3	2	3	6	9	4	5	3
d	14	32	12	7	16	7	16	5

Figure D12.3 Data for the Eight Job Tbar Problem

D12.4 This is an excerpt from a simulated annealing minimization problem. The seed schedule has a measure of 78. The four candidates for selection to the next step are 82, 84, 79, and 87. Determine the probability of being selected for each of these, using k/K = 0.5.

D12.5 Use the available seed schedule methods to obtain a good C_{max} for the $5/3/P/C_{max}$ problem in Figure D12.5. Discuss their effectiveness. All ready times are 0.

Job	P_1	P_2	P_3
1	8	6	3
2	8	5	4
3	2	7	1
4	4	5	3
5	4	11	4

Figure D12.5 Five Jobs, Three Machines

D12.6 Apply the genetic algorithm method to find a good solution to the C_{max} problem in Figure D12.6. Keep in mind that the method is not self-terminating, so you do not want to go on forever. How good is your solution? Note that you will have to calculate C_{max} many times, so be sure to do it automatically. All ready times are 0.

	Processing times				
Job	a_i	b_i	c_i	d_i	e_i
1	8	5	8	2	9
2	7	10	12	3	3
3	4	10	3	2	5
4	8	8	7	1	2
5	11	7	9	2	7
6	12	6	15	2	12
7	6	4	5	3	1
8	7	8	3	2	4

Figure D12.6 Data for the Genetic Algorithm Problem

D12.7 You are the production scheduler in a small company. On your way to work on Monday morning you have a flat tire! When you finally get to work, your enterprising floor manager has already started the shift by randomly selecting to start job 1 from the list in Figure D12.7 and is finishing it just as you walk in. You have only a minute to give her the next job. Your objective is to finish all jobs as soon as possible. Which one will you give her and why? There are eight machines in this flow shop and six jobs.

	M1	M2	M3	M4	M5	M6	M7	M8
Job1	10	15	12	3	17	22	5	11
Job2	11	12	5	5	5	6	6	2
Job3	5	4	6	7	4	6	9	7
Job4	8	8	8	8	6	9	7	15
Job5	9	5	22	31	4	5	7	8
Job6	7	7	4	5	11	15	6	5

Figure D12.7 Eight Machines, Six Jobs

D13.1 Use the longest processing time algorithm on the following problem. There are m machines and (2m + 1) jobs with

$$P_{2k-1} = P_{2k} = 2m - k \text{ for } i = 1, 2, \ldots, m \text{ and } p_{2m+1} = m$$

Show that the longest processing time schedule has $C_{max} = 4m - 1$, whereas the schedule found by using the list $(J_1, J_2, \ldots J_{m-1}, J_{2m-1}, J_{2m}, J_m, J_{m+1}, \ldots J_{2m-2}, J_{2m+1})$ has $C_{max} = 3m$. Then show that $R_A(I) = 4/3 - 1/3m$ is the best possible result for the longest processing time algorithm.

D13.2 Suggest a heuristic for parallel-machine problems where the objective is to minimize average flow time. Apply your heuristic to this 10-job, three-machine problem, and obtain the schedule and the average flow time.

Job	Processing Time	Job	Processing Time
1	4	6	7
2	5	7	1
3	6	8	2
4	3	9	4
5	3	10	8

Figure D13.2 Heuristic Parallel Machine Problem

D13.3 In a particular workshop, 10 jobs are run through three non-identical parallel machines. Their processing times are in Figure D13.3. Find a good solution for the minimum time that all the jobs can be completed.

Jobs	1	2	3	4	5	6	7	8	9	10	11	12	13
Machine 1	9	5	8	10	10	6	8	9	10	11	4	6	7
Machine 2	7	5	7	9	9	5	7	8	9	10	4	3	7
Machine 3	10	6	9	11	11	7	9	10	11	13	5	8	7

Figure D13.3 Data for Non-Identical Machines

D13.4 You have a small machine shop with three lathes, three grinders, and three vertical milling machines. Today you have seven jobs, all with the same technological constraint (Lathe–Grinder–Mill) and the following processing times:

You are interested in finishing as early as possible. How would you approach scheduling the day's work and why? What is the longest time it may take you to finish all the jobs (without doing the problem)?

Job	L	G	M
1	6	5	6
2	4	3	3
3	9	2	4
4	8	9	9
5	7	5	6
6	2	2	3
7	4	3	4

Figure D13.4 Machine Shop Processing Times

D14.1 In an operation where it is 1.5 times as expensive to be tardy as it is early, we are to determine a sequence of jobs that is optimal with the sum of early and tardy penalties. The processing times for 15 jobs are given in Figure D14.1. Give the sequence and the earliest due date that will be optimal.

Job	1	2	3	4	5	6	7	8	9	10	11	12	13	14	15
p	10	12	5	4	7	9	2	5	11	12	4	5	3	2	5

Figure D14.1 Early/Tardy Processing Times

D14.2 Using dynamic programming, complete the solution to the following four-job cyclical sequence dependent set up problem.

		Predecessor			
		1	2	3	4
Follower	1	-	15	10	20
	2	35	-	55	5
	3	30	40	-	25
	4	50	45	60	-

Figure D14.2 Sequence Dependent Set-Ups

D14.3 An oven can be set to three different temperatures for heat treating metal alloys: 1,000°F = A, 1,100°F = B, and 1,240°F = C. It can hold two jobs at any one time. We have 12 jobs to process, for which the temperature and processing time are given in Figure D14.3. We also have arrival dates for each and due dates. Our objective is to minimize the average tardiness. Develop a schedule that will attempt to do that and determine the average tardiness of your schedule.

	Processing time	Oven capacity
Job type A	3	2
Job type B	2	2
Job type C	2.5	2

Job	Type	Arrival	Due date
1	A	0	5
2	A	0	8
3	A	1	8
4	A	7	14
5	A	8	15
6	A	9	14
7	B	4	17
8	B	5	13
9	B	6	12
10	C	2	8
11	C	8	13
12	C	9	13

Figure D14.3 Oven Baking Times

D14.4 Determine the minimum optimal common due date for this 26-job one-machine problem when the penalty for tardiness is three times the penalty for earliness (Earliness penalty is 1) (4 points). What is the penalty for job 6?

Job	p	Job	p
1	8	14	15
2	7	15	5
3	5	16	3
4	9	17	6
5	3	18	2
6	4	19	9
7	3	20	3
8	8	21	8
9	4	22	12
10	7	23	9
11	12	24	11
12	7	25	18
13	8	26	13

Figure D14.4 Common Due Date Data

D14.5 A production facility has grouped its jobs by using common setups. An example of their production requirement for 14 jobs is given in Figure D14.5. The objective is to minimize a combination of maximum completions time and total tardiness, defined by the equation $0.25 \star C_{max} + 0.5 \star$(Total Tardiness). Construct one reasonable schedule that tends to minimize the total penalty and calculate the penalty. What is a lower bound for the maximum completion time?

Job	Type	Processing time	Arrival time	Due date
1	A	3	0	17
2	A	6	0	7
3	C	6	0	12
4	C	3	0	20
5	D	2	18	30
6	D	4	18	43
7	D	4	32	65
8	A	2	4	15
9	B	5	55	72
10	B	6	40	55
11	A	5	5	14
12	C	6	4	20
13	B	5	22	75
14	B	4	5	18

Setup times	
A	3
B	2
C	4
D	3

Figure D14.5 Grouped Jobs

D14.6 An operation can be represented by a single machine. A net present value with an annual discount rate of approximately 15% (0.3% weekly) is used to evaluate the schedules. They use four parameters (all to be discounted to the present time):

1. Receipt for value at the time of delivery at the later of completion or due date.
2. Cost of material at ready time.
3. Penalty for earliness for finished goods inventory held from completion to due date is 0.5% per week per value. Paid in whole at the time of delivery.
4. Cost of tardiness. Also paid at the time of delivery.

Given the data in Figure D14.6, find a good schedule and the corresponding lead time.

Job	Processing time	Due date	Value	Material	Tardy Penalty/weeks
1	5	8	10	2	1
2	2	20	15	5	1
3	7	17	5	1	1
4	4	18	8	3	1
5	3	9	12	4	1

Figure D14.6 Data for the Net Present Value Problem

D14.7 The injection of various impurities into silicon wafers in the production of integrated circuits is performed in chambers that must be kept closed and at constant conditions while the wafers are being processed. You cannot mix different wafers in the same batch. The facility in question operates 16 hours per day, five days a week. This schedule must be completed within the 16 hours of this day. Sequence these jobs so that they comply with being completed in the required time.

Job	Type	Arrival	Job	Type	Arrival
1	A	0	11	B	11
2	A	0	12	C	2
3	B	6	13	C	11
4	A	1	14	A	3
5	B	4	15	A	7
6	B	5	16	B	9
7	C	2	17	C	1
8	A	5	18	B	6
9	C	8	19	A	8
10	B	10	20	A	9

Chamber capacity	
A	4
B	4
C	4

Figure D14.7 Types, Arrivals, and Processing Times

D14.8 This is a one-machine problem with different penalties for each job for being tardy and penalties for being early. In addition, we are interested in minimizing the average flow time of the jobs. Describe a method of your own on how you would tackle this problem and then apply it to this small sample.

Job	1	2	3	4	5	6	7	8
Processing time	3	4	4	5	2	2	3	3
Due date	9	27	8	28	19	7	13	7
Early penalty	1	1	1	1	2	0	1	1
Tardy penalty	3	2	2	2	2	3	2	2

Figure D14.8 Different Early and Tardy Penalties

D15.1 Two parallel machines have exponentially distributed processing times. The average processing times for seven jobs are given in Figure D15.1. Determine the sequence on the two machines that will tend to minimize the expected maximum completion time.

Job	1	2	3	4	5	6	7
Average exponentially distributed processing time	6	9	8	12	12	4	6

Figure D15.1 Exponentially Distributed Processing Times

D15.2 The processing times in an $8/2/P/C_{max}$ are exponentially distributed. Assume that the buffer between machines is sufficiently large to effectively eliminate blocking. Find the sequence that will minimize the maximum completion time.

Job	1	2	3	4	5	6	7	8
P_{i1} ave.	2	3	4	1	3	4	5	2
P_{i2} ave.	3	2	1	4	4	3	3	4

Figure D15.2 Buffered Two Machine Problem

D15.3 Find an optimal schedule for the C_{max} measure, explain your method and state your reason for using it. How many optimal schedules are there for C_{max}? How did you calculate this? In an effort to balance the production line, one of the industrial engineers has suggested that the average processing times on the second machine be adjusted to match those on the first machine. How would this impact the optimal schedule? Why?

Job	Average P1	Average P2	Standard deviation of P1	Standard deviation of P2
1	2	3	2	3
2	3	2	3	2
3	4	3	4	3
4	2	3	2	3
5	2	3	2	3
6	3	2	3	2

Figure D15.3 Potentially Equal Processing Times

DA.1 A manufacturing operation uses 250,000 standard or earned labor hours per year to produce a variety of products. The other significant numbers are: \$18 million of non-material indirect overhead, \$1 million of material overhead, \$90 million of purchased direct material, and a direct labor cost of \$22/hour. You know that one of these products has 0.25 hours of labor and \$35 worth of material and has a total unit cost of \$59.50. What labor efficiency did this company assume to come up with the cost of \$59.50? If you had been given a labor efficiency of 95% and all overhead (non-material and material) was allocated to the products' labor rate, what would the total unit cost of this product be? Assuming a labor efficiency of 85% and that 10,000 of this product are produced in a year, for what percentage of the nonmaterial indirect overhead does this product pay? (Assume that material overhead is allocated to material.)

DA.2 Determine the unit cost of each of these products using the method discussed in Appendix A. Then compare the resulting costs when all the overhead is allocated to the labor.

Product	Labor hours per unit	Material cost per unit
Alpha	0.25	\$40
Beta	0.55	\$76
Gamma	0.12	\$170

Figure DA.2 Cost Information

Indirect overhead = \$120M
Material overhead = \$10.1M
Labor efficiency 85%
Total labor hours 1.9M
Direct labor rate \$27
Total material purchased \$410M
Number of products manufactured—approximately 200

DB.1 The time to complete a task in a project has been estimated to be completed in a most likely time of 8. The estimate for best possible time is 5, while for the most pessimistic time is 10. Using the beta distribution, the probability of completing the task by time 9 is 94%. What is the probability using a triangular distribution? A normal distribution (use the same average as you would for a triangular distribution)?

DB.2 Referring to the project network and data in Figure DB.2 and assuming a triangular distribution for each task:

1. Create an AON diagram based on most likely values, showing all start and finish times. What is the critical path when most likely durations are used and how long is it?
2. What is the probability of finishing by 24?

Task	Predecessor	Durations Best Possible	 Most Likely	 Worst Case
A	-	2	5	7
B	A	3	7	9
C	A	4	6	11
D	C	4	5	9
E	B	6	8	9
F	D	1	2	4
G	E,F	2	3	6

Figure DB.2 Project Network Data

Given the data for a project in the Figure DB.3, answer the following:

1. Draw an AOA network and an AON network.
2. Find the slack of each activity and the critical path.
3. Determine the probability of completing the project by time = 18 (use triangular distribution).
4. Use simulation to determine whether other paths could be critical (use triangular distribution).
5. Determine the relative criticality of each of the tasks.
6. How would your answers change if you used a beta distribution for the durations?

Initial Node	Ending Node	Optimistic time	Most Likely time	Worst Case time
A	B	3	6	9
A	C	2	4	7
C	B	1	3	6
C	D	3	3	3
C	E	2	2	8
B	D	0	0	6
B	E	2	5	9
D	F	4	4	10
B	F	5	8	12
D	E	1	1	1
E	F	2	4	7

Figure DB.3 Eleven Activity Project Data

DB.4 Given the project network data in Figure DB.4:

1. What is the critical path when deterministic durations are used? Show your calculations.
2. What is the most likely critical path when using triangular approximations of durations? Show your calculations.
3. What is the probability of completion by time 26?

		Durations			
Task	Predecessor	Deterministic	Best possible	Most Likely	Worst Case
A		5	4	5	7
B		5	3	5	8
C	A	6	3	6	7
D	B	3	2	3	6
E	D	8	5	8	10
F	C,D	9	3	9	11
G	A,B	3	2	3	4
H	E,F,G	6	4	6	9

Figure DB.4 Data for Determining a Critical Path

DB.5 A project has the precedences and processing times shown in Figure DB.5. Determine the probability of completing the project by time 40. Also show an AON or AOA chart.

Task	Predecessor	Best	Most likely	Worst
A	-	4	7	11
B	-	8	8	9
C	A	5	6	11
D	A,B	12	14	18
E	C,D	6	7	12
F	B,E	5	14	18
G	F	6	8	10
H	D	4	10	12

Figure DB.5 Precedences and Processing Times

DC.1 Show that, if v is the size of a problem when all the data are represented in binary and η is its size when the same listing convention is adopted but numbers are represented to the base 10, then $v \leq (\log_2 10)\eta$ and $\eta \leq v$.

DC.2 Show that if an n/m/A/B problem has size v then $v! \geq (n!)m$ (Hint: show $v \geq nm$.)

DC.3 Show that the polynomial $a_n v^n + a_{n-1} v^{n-1} + \ldots + a_0$ is $O(v^n)$.

DC.4 Show that the following algorithms have polynomial time complexity:

1. Johnson's for the $n/2/F/F_{max}$ problem;
2. Lawler's for the $n/1// \max_{i=1}^{n} \{\gamma_i(C_i)\}$problem.

BIBLIOGRAPHY

Baker, K. R. and Scudder, G. D. (1990) Sequencing with earliness and tardiness penalties: A review, *Operations Research*, Jan/Feb 1990, vol. 38, No. 1.

Baker, K. R. and Trietsch, D. (2009) *Principles of Sequencing and Scheduling*, John Wiley and Sons.

Blazewicz, J., Ecker, K.H., Schmidt, G. and Weglarz, J. (1994) *Scheduling in Computer and Manufacturing Systems*, 2nd Ed., Springer Verlag.

Brown, R. G. (1967) *Decision Rules for Inventory Management*, Holt, Rinehart and Winston.

Buffa, E. S. and Miller, J. G. (1979) *Production-Inventory Systems*, 3rd Ed., Irwin.

Chapman, S. N. (2006) *The Fundamentals of Production Planning and Control*, Pearson Prentice Hall.

Conway, R.W., Maxwell, W.L. and Miller, L.W. (1967) *Theory of Scheduling*, Addison-Wesley.

Dreyfus, S.E. and Law, A.M. (1977) *The Art and Theory of Dynamic Programming*, Academic Press.

Duzère-Péres, S. and Lasserre, J-B. (1994) *An Integrated Approach in Production Planning and Scheduling*.

Fogarty, D., Blackstone, J. and Hoffman, T. R. (1991) *Production and Inventory Management*, South-Western.

French, S. (1982) *Sequencing and Scheduling*, Ellis Horwood.

Hillier, F.S. and Lieberman, G. J. (1990) *Introduction to Operation Research*, McGraw-Hill.

Lawrence, K. D. and Zanakis, S. H. (1984) *Production Planning and Scheduling*, Industrial Engineering and Management Press.

Morton, T.E. and Pentico, D.W. (1993) *Heuristic Scheduling Systems*, Wiley.

Muth, J.F. and Thompson, G.L. Editors (1963) *Industrial Scheduling*, Prentice Hall.

Pinedo, M. (1995) *Scheduling—Theory, Algorithms, and Systems*, Prentice Hall.

Plossl, G. W. (1994) *Orlicky's Material Requirements Planning*, 2nd Ed., McGraw-Hill.

Ross, S. M. (2009) *Introduction to Probability and Statistics for Engineers and Scientists*, 4th Ed., Elsevier.

Silver, E., Pyke, D. and Peterson, R. (1998) *Inventory Management and Production Planning and Scheduling*, 3rd Ed., John Wiley.

Sule, D.R. (1997) *Industrial Scheduling*, PWS.

Vollman, T.E., Berry, W.L. and Whybark, D. C. (1997) *Manufacturing Planning and Control Systems*, 4th Ed., Irwin.

REFERENCES

Adolphson, D.L. (1977) Single machine job sequencing with precedence constraints. *SIAM J. Computing*, 6, 40–54.

Ashour, S. (1967) A decomposition approach for the machine scheduling problem. *It. J. Prod. Res*, 6, 109–122.

Ashour, S. (1970) A branch and bound algorithm for the flow-shop scheduling problem. *A.I.I.E. Trans*, 2, 172–176.

Baker, K.R. (1974) *Introduction to Sequencing and Scheduling*, John Wiley.

Baker, K.R. (1975) A comparative survey of flow-shop algorithms. *Ops. Res.*, 1, 62–67.

Baker, K.R. and Schrage, L.E. (1978) Finding an optimal sequence by dynamic programming: An extension to precedence-related tasks. *Ops. Res.*, 16, 111–120.

Baker, K. R. and Trietsch, D. (2009) *Principles of Sequencing and Scheduling*, John Wiley and Sons.

Bellman, R. (1957) *Dynamic Programming*, Princeton University Press.

Campbell, H.G., Dudek, R.A. and Smith, M.L. (1970) A heuristic algorithm for the n-job. m-machine sequencing problem. *Mgmt. Sci.*, 16, B630–B637.

Coffman, E.G. Jr., Ed. (1976) *Computer and Job-Shop Scheduling Theory*, John Wiley.

Conway, R.W., Maxwell, W.L. and Miller, L.W. (1967) *Theory of Scheduling*, Addison-Wesley.

Cook, S.A. (1971) The complexity of theorem proving procedures. Proceedings of the Third Annual ACM Symposium on the Theory of Computing. Association of Computing Machinery, New York, 151–158.

Corwin, B. I. and Esogbue, A.O. (1974) Two machine flow-shop scheduling problems with sequence dependent set-up times: A dynamic programming approach. *Nav. Res. Logist. Q.*, 21, 515–524.

Dannenbring, D.G. (1977) An evaluation of flow-shop sequencing heuristics. *Mgmt. Sci.*, 23, 1174–1182.

Dantzig, G.B. (1960) A machine-job scheduling model. *Mgmt. Sci.*, 6, 191–196.

Fisher, M.L. (1980) Worst case analysis of heuristic algorithms. *Mgmt. Sci.*, 26, 1–17.

Fisher, M.L. and Jaikumar, R. (1978) An algorithm for the space-shuttle scheduling problem. *Ops. Res.*, 26, 166–182.

Garey, M. R., Graham, R.L. and Johnson, D.S. (1978) Performance guarantees for scheduling algorithms. *Ops. Res.*, 26, 3–21.

Garey, M.R. and Johnson, D.S. (1979) *Computers and Intractability: A Guide to the Theory of NP-Completeness*, Freeman.

Garey, M.R., Johnson, D.S. and Sethi, R.R. (1976) The complexity of flow-shop and job shop scheduling. *Math. Ops. Res.*, 1, 117–129.

Gere, W.S. (1966) Heuristics in job-shop scheduling. *Mgmt. Sci.*, 13, 167–190.

Giffler, B. and Thompson, G.L. (1960) Algorithms for solving production scheduling problems. *Ops. Res.*, 8, 487–503.

Giglio, R. and Vagner, H. (1964) Approximate solution for the three-machine scheduling problem. *Ops. Res.*, 12, 305–324.

Goldratt, Eliyahu M. (1990) Theory of constraints.

Graham, R.L. (1969) Bounds on multiprocessing timing anomalies. *SIAM. J. Appl. Math.*, 17, 416–429.

Greenberg, H.H. (1968) A branch and bound solution to the general scheduling problem. *Ops. Res.*, 16, 353–361.

Gupta, J.N.D. and Dudek, R.A. (1971) An optimality criterion for flow-shop schedules. *A.I.I.E. Trans.*, 3, 199–205.

Held, M. and Karp, R.M. (1962) A dynamic programming approach to sequencing problems. *J. SIAM*, 10, 196–210.

Hillier, F.S. and Lieberman, G. J. (1990) *Introduction to Operation Research*, McGraw-Hill.

Ignall, E. and Schrage, L.E. (1965) Application of the branch and bound, technique to some now-shop problems. *Ops. Res.*, 13, 400–412.

Jeremiah, B., Lalchandani, A. and Schrage, L. (1964) Heuristic rules towards optimal scheduling. Research Report. Department of Industrial Engineering, Cornell University, Ithaca, NY.

Johnson, S.M. (1954) Optimal two- and three-stage production schedules with set-uptimes included. *Nav. Res. Logist. Q.*, 1, 61–68.

Keeney, R.L. and Raiffa, H. (1976) *Decisions with Multiple Objectives*, John Wiley.

Kelley, J. (1969) Critical path planning and scheduling: Mathematical basis. *Ops. Res.*, 9, 296–320.

King, J.R. and Spachis, A.S. (1980) Heuristics for flow-shop scheduling. *Int. J. Prod. Res.*, 18, 345–357.

Kohler, W.H. and Steiglitz, K. (1976) Enumerative and iterative computational approaches. In Coffman, Ed. (1976). 229–287.

Lageweg, B. J., Lenstra, J.K. and Rinnooy Kan, A.H.O. (1978) A general bounding scheme for the permutation flow-shop problem. *Ops. Res.*, 26, 53–67.

Lawler, E.L. (1973) Optimal sequencing of a single machine subject to precedence constraints. *Mgmt. Sci.*, 19, 544–546.

Lawler, E.L. and Moore, J.M. (1969) A functional equation and its application to resource allocation and sequencing problems. *Mgmt. Sci.*, 16, 77–84.

Lenstra, J.K. and Rinnooy Kan, A.H.G. (1979) Computational complexity of discrete optimization problems. *Ann. Discrete. Math.*, 4, 121–140.

Lockyer, K.G. (1969) *An Introduction to Critical Path Analysis*, Pitman.

Lomnicki, Z. (1965) A branch and bound algorithm for the exact solution of the three machine scheduling problem. *Ops. Res. Q.*, 16, 89–100.

Moore, J.M. (1968) An n-job, one machine sequencing algorithm for minimizing the number of late jobs. *Mgmt. Sci.*, 15, 102–109.

Palmer, D.S. (1965) Sequencing jobs through a multi-stage process in the minimum total time-a quick method of obtaining a near optimum. *Opl. Res. Q.*, 16, 101–107.

Park, C.S. (2015) *Contemporary Engineering Economics*, 6th Ed., Addison-Wesley.

Ptak, C. and Smith, C. (2011) *Orlicky's Material Requirements Planning.*

Rinnooy Kan, A.H.G., Lageweg, B. J. and Lenstra, J.K. (1975) Minimizing total costs in one machine scheduling. *Ops. Res.*, 23, 908–927.

Schrage, L. (1970) Solving resource-constrained network problems by implicit enumeration-non pre-emptive case. *Ops. Res.*, 18, 263–278.

Schrage, L. (1972) Solving resource-constrained network problems by implicit enumeration-preemptive case. *Ops. Res.*, 20, 668–677.

Schrage, L. and Baker, K.R. (1978) Dynamic programming solution of sequencing problems with precedence constraints. *Ops. Res.*, 26, 444–449.

Silver, E.A., Vidal, R.V. and De Werra, D. (1980) A tutorial on heuristic methods. *Eur. J. Opl., Res.*, 5, 153–162.

Silver, E., Pyke, D. and Peterson, R. (1998) *Inventory Management and Production Planning and Scheduling*, 3rd Ed., John Wiley.

Smith, M.L., Panwalker, S.S. and Dudek, R.A. (1976) Flow-shop sequencing problem with ordered processing time matrices: A general case. *Nav. Res. Logist. Q.*, 21, 481–486.

Smith, W.E. (1956) Various optimizers for single state production. *Nav. Res. Logist. Q.*, 3, 59–66.

Story, A.E. and Wagner, H.M. (1963) Computational experience with integer programming for job-shop scheduling. In Muth and Thompson, Eds. (1963). 207–219.

Sturm, L.B.J.M. (1970) A simple optimality proof of Moore's sequencing algorithm. *Mgmt. Sci.*, 17, BI16–BI18.

Van Wassenhove, L.N. and Baker, K.R. (1980) A bicriterion approach to time/cost trade-offs in sequencing. Paper presented at the 4th European Congress on Operational Research, Cambridge, England, July 22–25, 1980. Submitted to A.I.I.E. Trans.

Van Wassenhove, L.N. and Gelders, L.F. (1980) Solving a bicriterion scheduling problem. *European. J. Opl. Res.*, 4, 42–48.

Wagner, H.M. (1959) An integer programming model for machine scheduling. *Nav. Res. Logist. Q.*, 6, 131–140.

White, C.H. and Wilson, R.C. (1977) Sequence-dependent set-up times and job sequencing. *Int. J. Prod. Res.*, 15, 191–202.

White, D. J. (1969) *Dynamic Programming*, Oliver and Boyd, Edinburgh.

Wight, O. W. (1981) *Manufacturing Resource Planning: MRP II*, Unlocking America's Productivity Potential.

INDEX